Fairies

Genre Fiction and Film Companions

Series Editor: Simon Bacon

FAIRIES

A Companion

Edited by Lorna Piatti-Farnell and Simon Bacon

PETER LANG

Oxford - Berlin - Bruxelles - Chennai - Lausanne - New York

Bibliographic information published by the Deutsche Nationalbibliothek. The German National Library lists this publication in the German National Bibliography; detailed bibliographic data is available on the Internet at http://dnb.d-nb.de.

A catalogue record for this book is available from the British Library.

Library of Congress Cataloging-in-Publication Data

Names: Piatti-Farnell, Lorna, 1980 – editor. | Bacon, Simon, 1965 – editor.
Title: Fairies : a companion/Lorna Piatti-Farnell, Simon Bacon.
Description: Oxford ; New York : Peter Lang, 2025. | Series: Genre fiction and film companions, 2631–8725 ; vol no. 17 | Includes bibliographical references and index.
Identifiers: LCCN 2024062153 (print) | LCCN 2024062154 (ebook) | ISBN 9781800799370 (paperback) | ISBN 9781800799387 (ebook) | ISBN 9781800799394 (epub)
Subjects: LCSH: Fairies in popular culture. | LCGFT: Essays.
Classification: LCC P96.F313 F35 2025 (print) | LCC P96.F313 (ebook) | DDC 398.21—dc23/eng/20250203
LC record available at https://lccn.loc.gov/2024062153
LC ebook record available at https://lccn.loc.gov/2024062154

Cover image: Photo by Meritt Thomas on Unsplash.
Cover design by Peter Lang Group AG

ISSN 2631-8725
ISBN 978-1-80079-937-0 (print)
ISBN 978-1-80079-938-7 (ePDF)
ISBN 978-1-80079-939-4 (ePUB)
DOI 10.3726/b19901

© 2025 Peter Lang Group AG, Lausanne
Published by Peter Lang Ltd, Oxford, United Kingdom
info@peterlang.com – www.peterlang.com

Lorna Piatti-Farnell and Simon Bacon have asserted their right under the Copyright, Designs and Patents Act, 1988, to be identified as Editors of this Work.

All rights reserved.
All parts of this publication are protected by copyright.
Any utilisation outside the strict limits of the copyright law, without
the permission of the publisher, is forbidden and liable to prosecution. This applies in particular to reproductions, translations, microfilming, and storage and processing in electronic retrieval systems.

This publication has been peer reviewed.

Contents

Acknowledgements — xi

Lorna Piatti-Farnell and Simon Bacon
Introduction — 1

Maria Giakaniki
Flash Fiction: The Blue Fairy — 13

PART I Beginnings and Evolutions — 15

Lesley McLean
The Cottingley Fairies (Arthur Conan Doyle, 1920) – The Science of the Supernatural — 17

Abigayle Farrier
Changing the Changeling Trope (Patrick Doyle, 1875(?)–1888) — 27

Saga Bokne
"Don't Call Me a Fairy" (Various, 1964–2006) – Folkloresque Metacommentary in Contemporary Fairy Fictions — 37

Francesca Bihet
A.I. Artificial Intelligence (Steven Spielberg, 2001) – Electronic Parasites — 45

PART II Belonging and Otherness — 53

Angana Moitra
Queene of Phayries (Various, 1590–1602) – Tudor Fairies — 55

Joan Ormrod
Jonathan Strange and Mr Norrell (Susanna Clark, 2004) – Neo-19th-Century Fairies — 63

Nick Freeman
"The New Daughter" (John Connolly, 2004) – Fairies and Fatherhood — 73

Lorna Piatti-Farnell
Carnival Row (Travis Beacham and René Echevarria, 2019–Present) – Fairy Ecologies of Difference — 79

PART III Daughters, Mothers, and Godmothers — 89

Gemma Files
The Muses — 91

Jenny Wise and Lesley McLean
Peter Pan (Clyde Geronimi, 1953) – The Rise of Tinker Bell — 93

Blair Speakman and Nancy Johnson-Hunt
True Blood (Alan Ball, 2008–2014) – Sookie Stackhouse as a Plaything — 103

Contents   vii

Amy Harris
The Daisy Chain (Aisling Walsh, 2008) – Changelings and
Motherhood   111

Rebecca Wynne-Walsh
A Cinderella Story (Mark Rosman, 2004) – Evolution of the Fairy
Godmother   121

PART IV   By Any Other Name   131

Kirstin A. Mills
Quink   133

Jo Anna Burn
The Elves (Terry Pratchett, 1992–2003) – The Fairy Folk
of Discworld   135

Fernando Gabriel Pagnoni Berns
Don't Be Afraid of the Dark (John Newland, 1973) –
Mischievous Boggarts   143

Kristin Aubel
Rivers of London (Ben Aaronovitch, 2011–Present) –
Urban(ised) Fairies   153

PART V   Across Cultures   161

Clay Franklin Johnson
Poem: The Moonlight Meeting of Lord Ortho and Setareh   163

Shane Broderick
"The Otherworld" (Anon., c. 1200 BCE–Present) – Revenge of
the Sídhe   167

Muskan Dhandhi and Suman Sigroha
Raja Ka Sapna [A King's Dream] (Anon., n.d.) – Haryanvi Fairies   175

Amy Lee
Onmyōji (Yumemakura Baku, 2003) – Female Shikigami as
Japanese "Fairies"   181

Margaryta Golovchenko
Tecna (Iginio Straffi, 2004–Present) – Fairy of Modernity   191

PART VI Across Media   203

Kirstin A. Mills
Butterfly Magic   205

Allyson Wierenga
"The Sums That Came Right" (Edith Nesbit, 1901) –
The Arithmetic Fairy   207

Péter Kristóf Makai
Changeling: The Lost (White Wolf, 2007) – Supernatural Alterity   215

Catalina Millán Scheiding
Prikosnovénie (Frédéric Chaplain and Sabine Adélaïde,
1991–Present) – Dark Wave Fairies   223

Contents ix

Kirstin A. Mills
The Fairy-Land of Science (Arabella Buckley, 1878) –
Fairies and Science   233

PART VII   Environmental and Ecological   241

J. S. Mackley
"Sir Orfeo" (Anon., Late 13th/Early 14th Century) –
Medieval Fairies   243

Morgan Daimler
The Call (Peader O'Guilin, 2016) – Contemporising the Irish Sídhe   253

Sophia Lange
Magic: The Gathering (Richard Garfield, 1993–Present) –
Queen Oona, Environmental Protectress   263

Lesley Hawkes and Kelly Palmer
FernGully: The Last Rainforest (Bill Kroyer, 1992) –
Australian "Little People"   271

Gemma Files
Fairy of the Forest   281

Bibliography   283

Notes on Contributors   307

Index   317

Acknowledgements

Books are strangely collective efforts and not just about those that are in the final manuscript, but all those that were part of it at the various stages, as well as those that listened to thoughts, that gave advice, or even just patiently listened to grumps and complaints when things didn't quite happen as they should have. Without any of those people, all those tiny pushes, shoves, and reasons to carry on books just wouldn't get over the line. So to all of you, many, many thanks.

Of course, there are those that more obviously helped to get the book completed and so huge thanks to everyone that managed to get their chapters written, their proofs corrected, and send it all in. With them, we would also like to thank Laurel and the team at Peter Lang who have helped make the final book the beautiful thing it is. And we'd also like to thank the reviewers, the often unsung heroes of academic publishing, whose input has helped shape our arguments and motivate us and make this the best book it could be. Lastly we'd like to thank our respective partners and families, who are the reason for doing anything important in this world, and we'd like to thank each other, as it wouldn't have happened without either of us.

Lorna Piatti-Farnell and Simon Bacon

Introduction

Approaching Fairies: Taxonomy, Meanings, Evolutions

Stories about fairies and the fae have long populated the imagination of many cultures around the world. Shrouded in half-written chronicles and oral accounts, fairies are figures of mystery. Their definition, look, and practices shift and change from context to context. Fairies are often at the centre of tales of folklore across countries; they are entangled with narratives of magic and the supernatural, while also being intrinsically tied to the workings of the natural world. Indeed, while "fairies have not been proven to exist", many people in several parts of the world "believe that they are real" (Bane 2013: 1). Various terms are often used to refer to these mysterious creatures and, on occasions, are used interchangeably – from fairy to fae, and beyond – while also suggestively maintaining characteristics that are notoriously difficult to define. At times, they are perceived to be tall, beautiful, and ethereal, capturing the mystical aspects of life just beyond human reach. Other times, they take on the form of pixies and are thought of as little tricksters, who are also forever ready to play pranks on unsuspecting humans. Fairies may display completely unique anatomies, which only vaguely recall the humanoid shape: as the proud owners of rounded and glittery eyes, webbed feet, and even wings, fairies push the boundaries of social, cultural, and even anatomical acceptability. When it comes to fairy origins, definitions, descriptions, geographies, and general countenance, no "single theory" is the most "plausible" (Keightley 2023: 7) and several co-exist via forms of assumption and deduction, continuing to prove the cultural evasiveness of fairies as a central part of their historical endurance and relevance. A taxonomy of fairies, in the broader sense, is as difficult as it is fascinating, and our shifting categorisations often speak to how we as humans respond to the world, known and unknown. Fairies commonly stand as an embodiment of the otherworld and inhabit – as far as the human imagination goes – locations that are often perceived to be as secretive and distant, such as forests and rivers. Equally, fairies can exist as hidden and undiscovered presences in our everyday contexts, from gardens to kitchens; the fairies' elusive and capricious nature often speaks to their aloofness in relation to human matters. The "fairy world", as both a physical and a metaphorical location, is constructed of legends and cultural whispers that often lie beyond the limits of the rational imagination, while also being profoundly intertwined with what humans consider as equally whimsical and frightening. Entangled as they are with matters of folklore, memory, and, importantly, humanity, it is not surprising to see that fairies have also occupied a central role in our storytelling practices. Indeed, the historical problem with taxonomy, and the difficulty in providing an overarching definition of fairies – inclusive of sub-categories and transnational nuances – also provides the promise of innovation, especially as far as the 21st century is concerned. Adaptable and easy to mould to different contexts, fairies provide the ideal intersectional metaphor for capturing the evolving social, cultural, and political preoccupations that blend, mingle, and merge in the folds of changing historical backdrops. This adaptability is the starting point for this companion, as the volume openly aims to engage with the taxonomical, evolutionary, and representational rebelliousness of fairies, and their ability to both reflect and challenge the cultural certainties of the era they inhabit, representing changing ideas of gender, race, and ethnicity, as well mutating discourses of memory, belonging, and power. As daringly claimed by fairy scholar Richard Bullivant, as belief systems change and the world embraces cultural cycles of modernity and post-modernity, one thing about fairies remains certain: "They just refuse to go away" (2017: 2).

Fairy Lore, Folklore Studies, and the Fairy Imagination

Fairies have emerged as a steady presence in literature across the transnational landscape. Tales of fairies have been re-told in numerous contexts, building on central ideas about where they live, what they look like, and how they interact – if ever – with our human consciousness. Some geographical contexts and cultures have provided particularly fertile ground for the emergence of fairy tales; here, the British Isles, Scandinavia, and Germany are particularly worthy of note, and can be traced back centuries. For instance, some of the earliest written mentions of fairies in the English context occur in the "Anglo-Saxon charms against elf-shot"; equally, some of the "fairy ladies of medieval romances" may well have an "an origin as old", with examples such as Morgan Le Fey of Arthurian legend being based on a "mingling of Celtic and classical tradition" (Briggs 2002: 4). Fairies famously appeared in well-known Shakespeare's plays such as *A Midsummer Night's Dream* (1596), cementing their presence as an imaginative device for narrative representation. Indeed, in the literary context, the "fairy tradition" remains a recurrent part of representation. As Katharine Mary Briggs argues, "as tenuous and fragile as it is", that tradition is "still there" even in our contemporary moment, and "lingers on from generation to generation substantially unchanged" (2002: 3). Fairies and fairy-like forms commonly appear in classic fairy tales, written commonly between the 17th and 19th centuries, and which, perhaps as a result, became commonly remembered as both a folkloristic and literary genre that deals with all manner of magic, legends, and cautionary tales for children about conduct and behaviours. The 19th century proved particularly prolific in the production of fairy stories, which continued to explore the connections between literature and folklore. These narrative evocations were well suited to the Victorian fascination with the fairy world, situated as it was in the midst of cultural transformations, religious questioning, and a newly developing fascination with tales of the past (Silver 2000). This context provided a fruitful ground for the rediscovery of fairies as a storytelling presence and a much-debated topic of conversation – somewhat surprisingly – in historical, literary, and even scientific circles. In the late 1800s, acclaimed poet and scholar W. B. Yeats famously published a two-volume collection of *Irish Fairy Tales and Folklore*. Tales of fairies became entangled with discussions about both the natural world and our human impact, often mixing and mingling them with a preoccupation with the occult and the supernatural. The 19th-century fascination with fairies spread quickly from Europe to North America and beyond, soon incorporating long-standing tales of the fairy folk from different parts of the world, from Thailand to

Japan. Overall, however, Europe remains the most prolific continent for fairy lore, comprising a veritable compendium of legends, traditions, rituals, performances, and tales (Fay and Manwaring 2021).

In spite of the recurrent presence that fairies have maintained in our storytelling practices – including their evolution into recent media – academic interest in "fairy people" has been primarily sited in the field of folklore studies. Here, a large number of volumes have provided a critical overview of the place that many fairy incarnations – from pixies to the fae, and beyond – have occupied in our historical lore. Folklore and historical studies of the fairy world have been comprehensive and have often maintained a distinct encyclopaedic direction (Alexander 2014; Bane 2014; Cooper 2014; Daimler 2020; Guiley 2010; Hartland 2022; Henderson and Cowan 2001; Keightley 2023; Sugg 2018; Young and Houlbrook 2022). The focus in these studies has been primarily placed, somewhat unsurprisingly, on the European context, – with Ireland figuring prominently (Frost 2022; White 2005) – and has provided a picture of fairies as quasi-mythological creatures, sitting somewhere between oral tradition and religious beliefs. This is not to say, of course, that critical explorations of fairies around the world do not exist; indeed, Michelle Roehm McCann and Marianne Monson-Burton's *Finding Fairies* (2001) is a compelling study of fairy lore from six continents, even if the tone of the volume is more light-hearted than strictly scholarly. Making a more scholarly contribution to studies on fairy representation, Regina Buccola's volume *Fairies, Fractious Women, and the Old Faith: In Early Modern British Drama and Culture* (2006) is successful in showing how fairies have often functioned as veiled metaphors for the exploration of gender and class in the history of British theatre performances, and how these representations have both shaped and been shaped by cultural values.

It is perhaps unsurprising to see children's literature as one of the biggest exponents in its engagement with fairies, especially from the 20th century onwards. In similar terms, the genre of fantasy has capitalised greatly on representing fairies. It is difficult to forget the impact of fairies in Terry Pratchett's iconic *Discworld* series (1983–2015). Equally, it would not be too ambitious to identify the representation of Tinkerbell in J. M. Barrie's *Peter Pan* novels and stage plays (1902–1911) as one of the most well-known embodiments within the literary fairy world. Tinkerbell herself continues to maintain a legacy that, of course, goes well beyond her roots in the literary medium. Keen observers and fans would notice her recurrent presence in the Disney world, spanning multiple animated movies – including *Tinkerbell and the Lost Treasure* (2009) and *The Pirate Fairy* (2014), among others – and, of course, many lines of bestselling merchandise.

Indeed, fairies have also gained a noticeable importance in the 21st century, bringing with them an increased cultural focus on traditional beliefs and Indigenous identities. While the connection to the folkloristic and the literary remains strong – with the multiple reincarnations of Tinkerbell taking centre stage here – fairies have also found renewed life in modern and contemporary reimaginings. As Skye Alexander puts it, "it seems as if fairies are everywhere": "we're inundated with big-budget films, TV shows, and enough merchandise to fill every palace in fairyland" (2014: 7). Different interpretations of fairies – and more broadly, perhaps, the fae – have become a recurrent presence in popular culture. Film and television, as well as recent SVOD platforms such as Netflix and Amazon Prime, have provided a fertile arena for fairies to grow in influence and representation, especially considering their continued centrality in the cross-century genres of paranormal romance and fantasy. Fairies famously appeared as highly sexualised creatures in the popular television series *True Blood* (2008–2014), where their exploits were suggestively entangled with matters of greed, possession, and magic, and fairies themselves served as an embodiment – in a not-so-unexpected twist – of the darkest human desires. The fairy world is explored amply in recent examples of serialised narratives such as *Carnival Row* (2019–2021), where ethnically diverse "breeds" of the fae – from winged fairies, to fawns, pixies and beyond – mix and mingle with humans in a fictional 19th-century world. Fairies populate tales across genres and readerships, from Neil Gaiman's comic book series *The Sandman* (1989–present) to children's animated stories such as *Winx Club* (2012–present) and continue to captivate: their representations and meanings grow and expand, mixing ideas of magic with the inescapable nature of our tangible human lives. From the gender-swap production of *A Midsummer Night's Dream* (2020) to the portrayal of Billy Porter as a gender-neutral Fairy Godmother in *Cinderella* (2021), these re-envisionings give the old tropes of classic fairy texts new life.

Into the Evolving World of the Fae: This Book

The connection between fairies and tales of the imagination – often with a fantasy stance – is indeed unarguable, and yet the scholarly literature currently in existence has not shown a consistent desire to explore fairies not just as creatures of lore, but as central figures in the workings of the imagination, from literature to film, games, television, animation, and the broader popular culture landscape. While some important examples exist – such as Briggs' volume *The Fairies in Tradition and Literature* (2002) and Kristian Moen's *Film and Fairy Tales: The Birth of Modern Fantasy* (2013), a gap still exists in the critical analysis of fairies as popular representation entities. In answer to this, our companion aims to provide a broader view of the place occupied by fairies as an important part of cultural practice as well as culturally inscribed notions of magic and spirituality. Merging old lore with contemporary socio-cultural and socio-historical politics, the newly re-vamped fairies operate as central figures in the evolution of identity politics. Alongside this lies an obvious connection to the environment, where fairies become representative and protectors of the ecosystem, though not just as preservers of the past but as augurs of a future where humanity and the planet can survive together. In their multiple incarnations, fairies prove how the magical can be returned into the everyday. And in approaching the "fairy realm" as part of our collective imagination, we remain attuned to the "changing nature" of these continuously appealing creatures (Sugg 2018: 2), and we enquire into their ability to shift, mutate, and adapt to our evolving media and cultural contexts.

The companion will be divided in seven parts that will trace the fae from the enchanted isle out across the world, as they have become increasingly adept at representing non-normative identities and gender fluidity as well as the dark side of magic and (non)human nature. To highlight the importance of evolution and change to the fairy in the popular imagination – indeed, they are as mutable and adaptable as that other chameleon of pop culture, the vampire – each section in this collection trades a different thread of fae evolution, as they not only change in aspect and characteristics but change across mediums, as well.

The collection begins with "Flash Fiction: The Blue Fairy" by Maria Giakaniki, which contains something of a dreamlike description of the fae as feminine and docile – not unlike a classical muse – but it ends with them being released, to be whatever they might become, and that launches us into what will follow. What does follow is Part I, "Beginnings and Evolutions", which traces a more general sense of how fairies have been perceived, particularly in relation to technology and whether fairies have a place in the real world. The first chapter is "The Cottingley Fairies (Arthur Conan Doyle, 1920) – The Science of the Supernatural," by Lesley McLean, which considers the unlikely connection between the writer of Sherlock Holmes and fairies that actually speaks to a very particular kind of late Victorian/early 20th-century connection between science and the supernatural that was willing to accept the factual reality of the fae. Next is "Changing the Changeling Trope (Patrick Doyle, 1875(?)–1888)," by Abigayle Farrier that continues this blurring of the borders between folktale and reality and discusses the continually problematic history of the changeling, which still accounts for cases of fillicide in small, traditional communities. "Don't Call Me a Fairy" (Various, 1964–2006) – Folkloresque Metacommentary in Contemporary Fairies Fictions by Saga Bokne examines a modern updating of the fairies myth which uses the idea of the mockumentary, and so applies the conceit of reality to its main character who, not unlike a character of reality TV, is one who speaks its mind to purposely contradict traditional expectations. The final chapter in this section, "*A.I. Artificial Intelligence* (Steven Spielberg, 2001) – Electronic Parasites" by Francesca Bihet, centres on a possible technological future where even artificial humans need to believe in real fairies as a manifestation of a world beyond what can be seen and that the future is always a place of hope.

Part II, "Belonging and Otherness", describes the evolution of the fairy in creating a sense of national identity, but equally one that creates otherness and outsidership. It begins with "Queene of Phayries (Various, 1590–1602) – Tudor Fairies" by Angana Moitra, which takes us to the Tudor period, where fairies, while being scorned by the Church, became a means to embellish one's ancestry, as explicitly seen here in the case of Queen Elizabeth I herself who was styled as the Fairy Queen of England. "*Jonathan Strange and Mr Norrell* (Susanna Clark, 2004) – Neo-19th-century Fairies" by Joan Ormrod brings us closer to the present in a supernaturally tinged representation of the 19th century, where the alluring yet dangerous land of the fae represents the turmoil around England both then and in the present. The national changes to the familial and when an insider changes to an outsider in "'The New Daughter' (John Connolly, 2004) – Fairies and Fatherhood" by Nick Freeman. Here the changeling motif is inverted to throw light on the difficulties of fatherhood in raising daughters who are no longer who he thought they once were. Following this is "*Carnival Row* (Travis Beacham and René Echevarria, 2019–present) – Fairy Ecologies of Difference" by Lorna Piatti-Farnell, which examines how the body of the Fairy can be deployed as a signifier of difference and outsidership highlighting populist anxieties over immigration, and communal and familial belonging.

Part III, "Daughters, Mothers, and Godmothers", focuses on the evolution of the feminine within the representation of fairies with examples that both conform and defy expectations. It opens with an original artwork, "The Muses," by Gemma Files, before moving onto the chapter, "*Peter Pan* (Clyde Geronimi, 1953) – The Rise of Tinker Bell" by Jenny Wise and Lesley McLean, which skews the normative slightly as it charts the evolution of the hugely popular Disney Fairy who has always remained, and problematically so, all (normative) woman. Something of this continues in "*True Blood* (Alan Ball, 2008–2014) – Sookie Stackhouse as a Plaything" by Blair Speakman and Nancy Johnson-Hunt, where the young female fae, Sookie, lives a highly sexualised existence, though one which allows her to navigate through a male world on her own terms. Agency and difference becomes more problematic in "*The Daisy Chain* (Aisling Walsh, 2008) – Changelings and Motherhood" by Amy Harris. This chapter considers a tale from present-day Ireland where the continued belief in the idea of Fairy changelings sees a young girl vilified because of her inherent otherness to the village society around her. This part closes with "*A Cinderella Story* (Mark Rosman, 2004) – Evolution of the Fairy Godmother" by Rebecca Wynne-Walsh, which examines a more mature fairy figure in that of the Fairy Godmother, who is an increasingly queered figure that is represented as being beyond normative ideas of gender.

Part IV, "By Any Other Name", looks at the evolving nature of "folk" and entities that are often seen as fae, or fairy adjacent. It opens with an original artwork by Kirstin A. Mills, "Quink", which behind us in our way of examining fairy-like creatures. "The Elves (Terry Pratchett, 1992–2003) – The Fairy Folk of Discworld" by Jo Anna Burn, begins the chapters in this section and considers the Discworld novels that are largely devoid of fairies, but replace them with the elves, a stubbornly traditional and superior species who eventually show the possibilities of acceptance and cooperation. This is followed by *"Don't Be Afraid of the Dark* (John Newland, 1973) – Mischievous Boggarts" by Fernando Gabriel Pagnoni Berns, which complicates the evolution of the fae with some far less friendly "little people" who not only remain existentially opposed to the non-traditional, but equally against female liberation and agency. This fourth part closes with "*Rivers of London* (Ben Aaronovitch 2011–present) – Urban(ised) Fairies" by Kristin Aubel, which examines a recent fantasy franchise centred on an alternate universe where humans and a multitude of fairies creatures exist alongside each other within the urban landscape, which becomes a vehicle to examine contemporary problems of exclusion, immigration, class, and ethnicity.

Part V, "Across Cultures", looks at how fairies, and entities that display the same characteristics within the context they appear, are a feature of many cultures across the world. This begins with a poem, "The Moonlight Meeting of Lord Ortho and Setareh," by Clay Franklin Johnson, which features a very classical fairies figure but speaks to their transformative and cross-cultural nature. The chapter, " 'The Otherworld' (Anon., c.1200 BCE–present) – Revenge of the Sídhe" by Shane Broderick, discusses the evolution and contemporary representation of the Irish "other people" as being inherently alien to the world of humans and necessarily contrary. This is followed by "*Raja Ka Sapna* [A King's Dream] (Anon., n.d.) – Haryanvi Fairies" by Muskan Dhandhi and Suman Sigroha, which considers Haryanvi folk tales from the north of India that utilise the figure of the fairies, seemingly to express female agency and socially prohibited relationships. "*Onmyōji* (Yumemakura Baku, 2003) – Female Shikigami as Japanese 'Fairies' " by Amy Lee takes us to Japan and the fairies-like handmaidens of a nature entity that have been linked to the development of self-identity and individual agency within historical Japanese society. Margaryta Golovchenko ends the section with "Tecna (Iginio Straffi, 2004–present) – Fairy of Modernity", which focuses on a very Italian and "modern" interpretation of the scientific fairy who is both aspirational while allowing for a belief in the more than natural.

The next section is Part VI, "Across Media," which looks at how fairies have evolved across and in relation to certain kinds of media and starts with an original artwork, "Butterfly Magic" by Kirstin A. Mills. Following this is "'The Sums That Came Right' (Edith Nesbit, 1901) – The Arithmetic Fairy" by Allyson Wierenga, which discusses how at the height of the 19th-century fascination with fairies, they became something of a pedagogic tool in being associated with logic, reason, and mathematics. "*Changeling: The Lost* (White Wolf, 2007) – Supernatural Alterity" by Péter Kristóf Makai examines the range and characteristics of a whole set of types of fae in the popular tabletop role-playing game that reflect the diversity and complications of the contemporary world. Next is "Prikosnovénie (Frédéric Chaplain and Sabine Adélaïde, 1991–present) – Dark Wave Fairies" by Catalina Millán Scheiding, which examines a very pan-European and contemporary vision of fairy-culture as interpreted through the lens of dark alternative music, highlighting how the fae are not just about the lore of individual nations but can continue to embody transnational and global concerns over the environment and identity. Lastly, "*The Fairy-Land of Science* (Arabella Buckley, 1878) – Fairies and Science" by Kirstin A. Mills traces the development of the Victorian scientific fairy into the present day and the invisible forces at play in the universe.

The collection draws to a close in the final section, "Environmental and Ecological", which focuses on the evolving entanglement of fairies and the ecological, where they have increasingly become symbolic of a return to nature and a more holistic approach to the environment. The first chapter is "'Sir Orfeo' (Anon., late 13th/early 14th century) – Medieval Fairies" by J. S. Mackley, which looks at the introduction of the word "fairy" into the English language and some early examples in lore and poetry where they are part of the natural (folk) environment and, in particular, the anonymously authored "Sir Orfeo". Following this is Morgan Daimler's "*The Call* (Peader O'Guilin, 2016) – Contemporising the Irish Sídhe" considers a more recent urban fantasy text that utilises elements of Irish folklore, highlighting the issue of adhering to, or ignoring, integrity to the original stories and the land they come from, and the complications around cultural appropriation. "*Magic: The Gathering* (Richard Garfield, 1993–present) – Queen Oona, Environmental Protectress" by Sophia Lange begins this section with a consideration of the giant female flower-fairy from the *Magic: The Gathering* franchise, highlighting the continued ubiquity of the fairy and its connection to the natural environment. In closing, we move from the American West Coast to Australia, and Lesley Hawkes and Kelly Palmer's chapter "*FernGully: The Last Rainforest* (Bill Kroyer, 1992) – Australian 'Little People'" considers a rather transnational vision of eco-fairies, specifically with an Antipodean flavour, though one that is not without its own complications. As something of an endnote, or final musical flourish, there is a final original artwork, "Fairy of the Forest" by Gemma Files, suggesting an impish figure that might be angelic, but could be malevolent, but definitely and defiantly not human. It looks directly at us with a glint of determination in its eyes, as it will not be turned from its cause no matter what we do.

This companion thus provides a distinctive and unparalleled view of the ubiquity of fairies and fairy culture across the world, in its many expressions and manifestations. As well as providing historical perspectives on our belief in fairies, the companion also uniquely suggests their ongoing evolution and importance to culture in the 21st century and our possible futures (together). With the inclusion of original artworks, poems, and flash fiction, this collection provides an invaluable and indispensable resource for scholars and non-academics alike who want to understand the richness of our entanglement with fairies and what they say about our past, our present, and, more importantly, our need for the magical in the future.

Maria Giakaniki

Flash Fiction: The Blue Fairy

The Blue Fairy looks like a common woman.

Around her there is a circle of little girls wearing white dresses.
They all have hair of the same length that reach their shoulders, and they wear white bows.

One by one, they start reciting from a book.
When they finish, they start singing the same fragment slowly, as in a ritual.

Then they come out to the garden in a row.

They move languidly, the air is dim through the strong sunlight and there are no flowers they can pick.

The only flowers of the garden have very thick stems and they are very high so they cannot touch them.

When they re-enter the room, the Fairy is absent.

They sit around the table and cut a cake in slices of the same size.

The hours pass.

They paint each other's portraits using chalks of various colours.

It has now become dark.
One by one, they fall asleep on the table and the floor.

*

In the morning, the Fairy enters the room.
She wears a tight black ribbon around her neck.

With her little scissors, she cuts a lock of hair from each of the sleeping girls and she puts them into a little black box.

*

The little girls start slowly, and with great difficulty, waking up again, but no-one can tell their age now or who they are.

And the Fairy is not there – they will be on their own from now on.

Part I
Beginnings and Evolutions

Lesley McLean

The Cottingley Fairies
(Arthur Conan Doyle, 1920)

Introduction

How did Arthur Conan Doyle, creator of Sherlock Holmes – the much-loved fictional detective famed for his skills in observation, logical analysis, and supposed deductive reasoning – come to defend the existence of fairies? Beginning with a short article published in the 1920 Christmas issue of *The Strand Magazine* entitled "Fairies Photographed" (1920), Doyle stunned the British public with photographs of fairies taken three years previously by 16-year-old Elsie Wright and her 10-year-old cousin Francis Griffiths, who were residents of the small English village of Cottingley in North Yorkshire at the time. The girls captured close-up shots of delicately winged, tiny humanoid creatures they claimed to have befriended in woodlands behind their house. Doyle's reasons for defending the Cottingley Fairies, as they later came to be known, appear connected to his commitment to spiritualism, and the aim here is to explore the history of the Cottingley Fairies from the perspective of Doyle's defence of the paranormal. For Doyle, belief in the existence of fairies alongside the kind of scientific rationalism so aptly represented by his character Sherlock Holmes held less inconsistency than one might imagine; the same held true for his views on spiritualism. Put another way, it would appear that – to his mind at least – upon investigation of the evidence, belief in fairy-folk and spiritualist phenomena was both reasonable and scientifically rational. Doyle's involvement in the Cottingley Fairy saga, much like the event itself, has fascinated historians, folklorists, and the general public ever since.

The Case of the Cottingley Fairies

In several *Strand Magazine* articles, including the aforementioned "Fairies Photographed" (1920) and the later book-length account entitled *The Coming of the Fairies* (1922), Doyle presented the case as he saw it, inviting readers to assess the evidence and to draw their own conclusions. Common Holmesian terms like "proof", "evidence", "investigation", "common sense", "facts", and "truth" were invoked alongside those the fictional detective would have associated with the credulous, such as "clairvoyants", "infinite vibrations", "psychic spectacles", "psychic powers", "spirits", and "ectoplasmic images' (Doyle 1922: passim). At the heart of Doyle's investigation were Elsie's and Frances' photographs, which he claimed would bring about a profound shift in human thought if proven authentic (Doyle 1922: 18). The most famous of the photographs, entitled "Alice and the Fairies" (see Figure 1), featured Frances (whom Doyle had given the pseudonym Alice Carpenter) surrounded by four small fairies, beautifully dressed, hair coiffed, joyously dancing and flying about her person (Doyle 1920: 465). In another photograph, "Iris and the Dancing Gnome" (see Figure 2), Elsie (whom Doyle have given the pseudonym of Iris Carpenter) sits quietly on the grass, smiling, with her arms in her lap beckoning a diminutive, winged creature towards her (Doyle 1920: 462). How these family photographs became supposed photographic evidence is well-captured in Doyle's writings on the topic; much less so, his inferences inviting belief in the existence of fairy-folk. Yet, that was the position Doyle himself had come to accept by publication of the book.

How the matter of the Cottingley Fairies arose for Doyle was very much informed by the investigative work of paranormal enthusiast and prominent theosophist Edward L. Gardner. In 1920, Polly Wright, Elsie's mother, attended a lecture organised by the local Theosophical Society in which the topic of fairies was raised (Doyle 1922: 26). With mention of the photographs, she was invited to send them to the lecturer who in turn sent them to Gardner. Gardner wasted little time sourcing the original plates from which the positives had been developed, much to the surprise of Elsie's father, Arthur Wright, who had been sceptical of their content from the outset. It was Arthur's Midg, quarter-plate box camera Elsie had borrowed to take the photographs that now so excited

Figure 1. "Alice and the Fairies" 1917. Image in the public domain.

Figure 2. "Iris and the Dancing Gnome" 1917. Image in the public domain.

Gardner (Doyle 1922: 35). It had been Arthur, too, who had developed the negatives in his amateur dark room at the Wright family home. Invoking not a little rhetorical flourish, Doyle (1922: 60) recounts Elsie exclaiming to her younger cousin during the development process, "Oh, Alice, Alice, the fairies are on the plate – they are on the plate!"

Gardner initially took the plates to an expert named Harold Snelling who determined the photographs to be authentic images of fairies, or more specifically, who determined that trickery in the photographic process had not been employed (Doyle 1920: 464–5). Gardner, and later Doyle, took Snelling's report to mean endorsement of the reality of the fairies and went so far as to have Snelling touch-up the negatives to create better prints from which to examine the content (Doyle 1920: 465). Other experts were consulted, including several working for Kodak; however, they refused to testify to the genuineness of the photographs, despite, according to Doyle and Gardner, finding no flaws.

Under the patronage of Gardner and then Doyle, Elsie and Frances went on to take several more photographs of fairies, albeit with equipment supplied by Gardner, including secretly marked plates (Doyle 1922: 56). These and other photos of the woods and the waterfall where the fairies were purported to live were variously published with Doyle as sole author. The *Coming of the Fairies* (1922), while largely a rehash of previously published *Strand* articles, nevertheless crystallised for readers three key lines of defence in favour of the truth of fairy existence. The first concerned the authenticity of the photographs; the second concerned the status of the photographers themselves; and the third focused on the integrity of the family involved. For both men, it was inconceivable that two children of the "artisan class", who lived in a "quaint, old-world village in Yorkshire", had acquired enough skill and knowledge to deceive not only the experts, but the two men themselves (Doyle 1922: 18, 46). Later learning that Elsie could draw and had briefly worked for a photographer did little to raise their suspicions. Moreover, all members of the family were deemed above reproach. For example, Arthur Wright, they claimed, had exemplified both integrity and honesty in all dealings (Doyle 1922: 26), and while confirming the details of Elsie's account and accepting the girls' statements, he had nevertheless continued to voice scepticism regarding the existence of all things paranormal throughout the whole affair.

Reception and Revelation

It's fair to say reception of Doyle's and Gardner's findings in the case of the Cottingley Fairies caused great interest in the 1920s, with responses ranging from the supportive to the sceptical to the highly critical. In Chapter 3 of *The Coming of the Fairies*, Doyle reproduced in total a number of responses, particularly those that were quite critical. An entire article from the *Westminster Gazette* was reproduced; so too, pieces from the *Birmingham Weekly Post* and the *John o'London*, all of which challenged the authenticity of the photographs. The latter ended amusingly with the author, Mr Maurice Hewlett, stating that "knowing children, and knowing that Sir Arthur Conan Doyle has legs, I decide that the Miss Carpenters have pulled one of them" (quoted in Doyle 1922: 79).

The Cottingley Fairy case, and Doyle's involvement within it, never quite disappeared from public consciousness in the years after the event (see especially Owen 1994: n.1, 81). Indeed, the 1960s and 1970s saw an upswing in media attention, including write-ups in a range of magazine and newspaper articles as well as TV interviews and documentaries revisiting the events and the people involved. In 1965, a reporter from the *Daily Express* in England managed to track down Elsie, who like Frances, had long since married and moved away. In denying they were fakes, she nevertheless described the photos as "pictures of figments of our imagination, Frances and mine, and leave it at that" (cited in Smith 1991: 393). Some, of course, refused to "leave it at that". James "The Amazing" Randi, a stage magician and sceptic who sought to debunk claims concerning paranormal phenomena conducted his own investigation into the affair; so too photographic expert and editor of the *British Journal of Photograph*, Geoffrey Crawley (Smith 1991: 394–5). Randi argued that the photos were of cardboard cutouts of fairies, even going so far as to identify the book from which the illustrations were likely taken: *Princess Mary's Gift Book* (1914: 104; see Figure 3). Crawley (1982–1983) also concluded that the photographs were inauthentic; however, his was an investigation conducted and published over several years, focusing as much on the people and complex nature of relationships involved as on the photographic equipment used and their associated processes.

Figure 3. Illustration of Fairies in *Princess Mary's Gift Book* (1914), Image in the public domain.

In 1983, news broke of Elsie's and Frances' admission to having faked the photographs; that the images were indeed not those of real fairies, but of cardboard cut-outs held up by hatpins. It was not Crawley who initially broke the news; instead, the hoax was revealed in *The Times of London* newspaper with the headline: "Cottingley fairies a fake, woman says" and later "Secrets of two famous hoaxers" (Hewson 1983: 3). Elsie nevertheless sent Crawley a letter thanking him for his understanding of the difficult situation she and Frances had found themselves in. What had started as a practical joke perpetrated by children soon morphed into a public debate amongst eminent adults invested in the truth of fairy existence. Aware of, particularly, Conan Doyle's investment in the answers Elsie expressed much sympathy for the author. She knew of the mocking criticisms he had received for his belief "in our fairies", as well as his commitment to spiritualism. She had also heard of the death of his son, Kingsley, during the First World War, suggesting "the poor man was probably trying to comfort himself with unworthy things" (NSNM: Elsie's Letter).

Fairies, Spiritualism, and Victorian Science

Doyle had indeed suffered tremendous loss during the Great War, and the inference that he, like so many others, had sought solace in beliefs and practices they might otherwise have ignored or found wanting, is not unreasonable, but nevertheless inaccurate. Spiritualism, which promotes the view that spirits of the deceased survive physical death and are capable of communicating with the living through human mediums (and vice versa), had informed Doyle's life in one way or another long before the war. As a young man in the 1880s, he had joined the Society for Psychical Research (SPR) whereby members sought to "examine without prejudice or prepossession and in a scientific spirit those faculties of man, real or supposed which appear to be inexplicable in terms of any generalized hypotheses" (quoted in Wynne 1998: 387). Over the many years that Doyle was involved with the SPR, his investigations remained steadfastly focused on the reality and authenticity of spiritualist phenomena and mediumship, culminating in his conversion to the movement in 1916, one year prior to Elsie's and Frances' first fairy shots.

> I might have drifted on for my whole life as a psychical Researcher, showing a sympathetic, but more or less dilettante attitude towards the whole subject ... I seemed suddenly to see that this subject with which I had so long dallied was not merely a study of a force outside the rules of science, but that it was really something tremendous, a breaking down of the walls between two worlds, a direct undeniable message from beyond, a call of hope and of guidance to the human race at the time of its deepest affliction. (Doyle 1918: 38)

It was not only Doyle's promotion of spiritualism that largely served as the backdrop to the Cottingley Fairies photographs, but it was also the key to understanding his position regarding the paranormal more generally.

In the preface to *The Coming of the Fairies* Doyle warned readers against making too much of the connection between fairies and spiritualist doctrine (1922: 3–4). The latter, he said, did not depend upon the objective reality of the former for its integrity or truth-value; the two were not mutually inclusive. However, he did not think them mutually exclusive either, claiming that recognition of fairy existence would

> jolt the material twentieth-century mind out of its heavy ruts in the mud, and ... make it admit that there is a glamour and a mystery to life. Having discovered this, the world

will not find it so difficult to accept that spiritual message supported by physical facts which has already been so convincingly put before it (Doyle 1922: 41).

What is evident from this passage is Doyle's positioning of himself and his work within the science of the age, whilst nevertheless criticising its perceived reliance on the principles of materialism to the exclusion of all other explanatory models of existence. Victorian science, he said, would leave "the world hard and clean and bare, like a landscape on the moon", shedding little light on the mysteries of life that loom in the shadows, difficult to ignore (Doyle 1922: 117). The paranormal phenomena associated with both spiritualism and the Cottingley event "represented for him that 'glamour and a mystery to life' which proved the limitations of scientific knowledge and the limits of materialism" (Owen 1994: 69).

These mysteries were nevertheless captured in the language of new and emerging technologies of late 19th and early 20th century. Where Victorian scientists drew on the language of light waves, frequencies, wavelengths, and electromagnetic fields to explain the inner workings of wireless radios and visual mechanics for example (Branford 2011), Doyle similarly invoked the language to hypothesise the existence of fairies, suggesting them "only separated from ourselves by some difference of vibrations" (Doyle 1922: 1). Tuning ourselves to their vibrational frequency, or them to ours, is within the realm of scientific possibility. If objects exist, he said, and "the inventive power of the human brain is turned upon the problem, it is likely that some sort of psychic spectacles, inconceivable to us at the moment, will be invented, and that we shall all be able to adapt ourselves to the new conditions" (Doyle 1922: 1). Such was the nature of Doyle's speculation at the outset of *The Coming of the Fairies*.

Conclusion

Drawing together the discussion thus far, and employing speculation of a different form: what would (Doyle as) Sherlock Holmes have made of the Cottingley Fairies? The answer lies in Holmes' response to Dr Watson in the "The Adventures of the Devil's Foot" (1910). Holmes exclaims: 'I take it, in the first place, that neither of us is prepared to admit diabolical intrusions into the affairs of men' (Doyle 2003: 475). The comparison is clear: where Holmes, the super-rationalist, held no truck with fairies, spirits, and goblins, a credulous Doyle openly promoted their existence.

Yet, where mystery did not revert to the paranormal in the Holmesian context, it did not revert to a world "hard, clean, and bare", either. His "science of observation", wrote Michael Saler, "demonstrated that profane reality could be no less mysterious or alluring than the supernatural realm; the material world was laden with occult significance, which could be revealed to those with an observant eye and logical outlook" (Saler 2003: 614). While Doyle admitted "diabolical intrusions" and his character did not, the latter was not the hard materialist he seemed at first glance, nor was Doyle necessarily the credulous romantic he appeared to be.

Perhaps, however, the final word should go to the women whose photographs lay at the heart of the fairy debate. Whilst both Elsie and Frances admitted to having faked the photographs, Frances, unlike Elsie, insisted that fairies were real and that she had seen them as a child. Indeed, according to Frances, the fifth and final photograph entitled "Fairies and Their Sun-Bath" captured real fairies; the photograph was not a fake (Smith 1991: 399; see Figure 4). Resembling the Doyle/Holmes situation: Elsie admitted no paranormal intrusions, whereas the (final) photograph validated Frances' belief in fairies.

Figure 4. The Fifth Photograph: "Fairies and Their Sun-Bath" 1920. Image in the public domain.

Abigayle Farrier

Changing the Changeling Trope (Patrick Doyle, 1875(?)–1888)

On 31 January 1888, the news of Patrick Doyle's death hit the press for the first time (*Kerry Sentinel* 31 Jan. 1888: 2). Was his death a murder, a tragic accident, or a terrible misunderstanding? Over the next several months, the story that unfolded would be contradictory, convoluted, and centre on a deeply rooted belief in fairies and changelings. These beliefs would have fatal consequences, all centring on the critical question: can fairylore be a rationale for filicide? And, for a modern-day audience, how has the changeling legend changed since the 19th century?

A History of Changelings and Changeling Killings

The changeling phenomenon extends back to the Middle Ages, but became extremely popular in the Victorian era, after the publication of collections of folkloric changeling stories (Munro 1997: 251). These stories popularised the idea of an individual who had been replaced with a fairy substitute and needed to be rescued from the fairy realm. Children were the most commonly seen victims in these cases, and their parents were forced to trick the fairy changeling by way of some sort of ruse, typically involving fire or iron, in order to banish the fairy and retrieve their child. Sometimes these attempts were successful – usually the instances seen in fairytales – and sometimes they were not – these were the cases seen more often in real life. The concept of changelings goes far beyond fairytales, however: the tradition of fairy changelings offered parents a way to provide an explanation for their child's mysterious illness, death, disability, or disappearance. The way changelings were presented, in oral tradition and in artistic renditions, suggests that the idea of a changeling was a coded way to discuss disability and illness, both physical and mental (Bourke 1995).

In the above image of a changeling and its mother (see Figure 5), note the physical appearance of the child changeling: the colour of its skin compared to the mother's, the deformation of its limbs, and the aged and wizened appearance of its face. All of these attributes that were typical of changelings were also signs of illness or disability (Silver 2000: 60). Without the proper ways to diagnose or cope with these realities, the idea of a fairy abducting a healthy child and replacing it with an ill one was often easier for parents and loved ones to deal with.

While this is no new observation – countless scholars have noted the similarities between modern medical understandings of disabilities and childhood failure to thrive with the changeling legend – a lesser explored reality is that of "changeling" killings. The concept of changeling killings, or the deaths of individuals suspected to be fairy changelings, may not itself be an uncommon one; in Irish fiction and folklore, and even in Victorian newspapers, there are numerous stories of women being abducted by fairies and replaced by a changeling. Historical events, however, offer a different angle from which to approach the phenomenon of the fae's influence on familial relationships. The tragic reality of the deeply ingrained belief in the changeling legend is that many families harmed their relatives, believing them to be a fairy rather than a flesh-and-blood family member. There are notable historical events of women, such as Bridget Cleary, being abused or murdered because they were suspected to be a changeling (Bourke 1995). These fairy abduction cases were an object of public fascination, and throughout the 19th century, a very living belief in "changelingism" was documented in numerous publications (Bourke 1995; Silver 2000; Vyse 2018). A common thread in this discourse is that women and children were almost universally the victim in these cases of fairy abductions. This is certainly the case with Bridget Cleary, whose torture and death at the hands of her husband is renowned for being the most well-known killing rationalised by a belief in fairylore (Bourke 1995). It has been studied by numerous scholars, notably Angela Bourke, and has been the topic of several articles and books (Bone

Figure 5. The Changeling by P. J. Lynch. Reproduced with the permission of the artist.

2022; McGrath 1982; Scheible 2022). Even at the time of its occurrence, Bridget Cleary's death was well-known and well-documented in popular press of the time, both in Britain and Ireland.

Patrick Doyle's death, however, though occurring eight years earlier and being perhaps as brutal as Cleary's, has received little scholarly attention. Likewise, though Patrick's case received extensive immediate media attention in Ireland, it was a short-lived media sensation and quickly disappeared from the press after the sentencing of those involved. How could this be? Patrick's case had several of the same noteworthy elements that Cleary's did: he died in the presence of numerous family members, suffered a brutal death, and was suspected of being a fairy changeling. Perhaps the answer lies in the details of Patrick's death and position within his family.

Newspaper Accounts of Patrick's Death

The *Kerry Sentinel*, an Irish newspaper, was the first newspaper to publish Patrick's death. On 31 January 1888, the *Sentinel* published that "an idiot boy, aged thirteen" had been found dead in the Doyle's front yard (*Kerry Sentinel*, 31 Jan. 1888: 2). He was in a "half-nude condition … the face mangled and torn after a sow which had evidently been feeding on it for some time". The article begins with the assumption that Patrick has died at the hands of his family members, but concludes the article by saying, "the boy was in a very weak state for sometime hitherto, and he is said to have been so far gone last week that it was reported he was dead, and the people congregated for the funeral". The timeframe of Patrick's death is thus thrown into question, but more importantly, the *Sentinel* goes on to question the cause of death, writing, "If the boy was in such a delicate state of health it is quite possible that had he been thrown into the yard in the state of semi-nudity in which the body was found, he would have died from exposure without any violence having been used towards him" (*Kerry Sentinel*, 31 Jan. 1888: 2.).

Over the next few days, newspapers across Ireland proceeded to present a sensationalised version of the originally reported events, particularly of Patrick's body. One day after the *Sentinel*'s account was published, *The Irish Examiner* described Patrick as a "a mangled corpse entirely nude, and stretched upon a wretched litter in an outhouse". They went on to claim that "the poor boy ... was killed in the maniac fury by which the whole family was seized" (The *Irish Examiner*, 1 Feb. 1888: 3). After this narrative was published, the idea was generally accepted, and all following publications operated under the assumption that Patrick died at the hands of his family members. Three days later, the *Weekly Irish Times* wrote that "stretched in a pool of blood was the mangled and partially nude corpse of Patrick Doyle". Patrick's family is also dramatised by the *Weekly Irish Times*, described as "raving lunatics" behaving in a "demoniacal manner," while "violently resist[ing] and struggl[ing] with the constabulary". Though the *Weekly Irish Times* admits that Patrick was a "little delicate imbecile" and that "it had been rumored that he had died", they argue that "there seems scarcely any doubt that the boy was killed by his parents and their children when struck by this extraordinary fit of insanity" (*Weekly Irish Times*, 4 Feb. 1888: 6.).

Though the newspaper accounts agree on little, from the state of Patrick's clothing to the location of his body, they do have two commonalities. First, every newspaper report refers to Patrick's disabilities. Some accounts allege that he has mental disabilities, but every account mentions a physical disability. The *Kerry Sentinel* describes Patrick as being "afflicted both physically and mentally", coupled with the descriptor "idiot", while The *Irish Examiner* goes further, detailing Patrick's disabilities (The *Kerry Sentinel*, 31 Jan. 1888: 2; The *Irish Examiner* 1 Feb. 1888: 3). According to The *Irish Examiner*, Patrick was a "miserable specimen of humanity, bereft of the power of speech and hearing, and an idiot subject to periodical fits of epilepsy" (The *Irish Examiner*, 1 Feb. 1888: 3). The *Kerry Sentinel* records that Patrick and his younger siblings are described by their neighbors as being "half-natural" (The *Kerry Sentinel*, 31 Jan. 1888: 2). The claim that Patrick's younger siblings are "half-natural" or intellectually disabled is later disproved, as one of the star witnesses in the trial is Denis Doyle, Patrick's younger brother, whose intelligence is praised by the court (The *Kerry Sentinel*, 3 Feb. 1888: 4). However, it is possible that genetics played a role in Patrick's condition, as the *Examiner* claims that Patrick's cousins were "mostly all deaf and dumb, or suffering under some other physical or mental infirmity. Four or five of the family died young, and of those still surviving, half are afflicted one way or another" (The *Irish Examiner*, 1 Feb. 1888: 3).

Neurological Diseases and Fairy Legends

On 16 July 1887, just a few months before Patrick's death, a "deaf mute" Patrick Doyle was arrested with his brother, Michael Jr, on the charges of sheep-stealing (*Kerry Evening Post*, 16 July 1887: 3). Based on this article, as recently as July of the previous year, Patrick was physically healthy enough to wrangle more than fifteen adult sheep; his only noted health concern was hearing loss. This instance is important to note, because it makes it apparent that sometime between July 1887 and January 1888, Patrick experienced a rapid physical decline, changing from being healthy and strong enough to steal sheep to languishing near death for over two weeks. His family attributed this to fairies, believing Patrick to be a changeling. There may, however, be a medical explanation for Patrick's decline.

Though it is impossible to definitively diagnose Patrick, he exhibits many symptoms of a neurodegenerative disease, such as Batten disease. The National Institute of Neurological Disorders and Stroke records that

> common symptoms for most of the forms include vision loss, seizures, delay and eventual loss of skills previously acquired, dementia, and abnormal movements. As the disease progresses, children may develop one or more symptoms including personality and behavior changes, clumsiness, learning difficulties, poor concentration, confusion, anxiety, difficulty sleeping, involuntary movements, and slow movement. Over time, affected children may suffer from worsening seizures and progressive loss of language, speech, intellectual abilities (dementia), and motor skills. Eventually, children with Batten disease become blind, wheelchair bound, bedridden, unable to communicate, and lose all cognitive functions. (U.S. Department of Health and Human Services 2018: 1–2)

Patrick presents with the majority of these symptoms, particularly at the end of his life. As the disease progresses, the symptoms become more and more prominent, likely causing Patrick's incapacitation and inability to feed himself, both of which are noted in Denis Doyle's testimony (The *Kerry Sentinel*,

3 Feb. 1888: 4). These symptoms are, of course, not limited to Batten disease – there are a number of adolescent-onset neurodegenerative diseases that have similar symptoms, and the specific neurodegenerative disease is less important than its larger social implications.

Fairylore and Families

Though it has been repeatedly argued that fairylore offers a way to discuss otherwise unfathomable situations, the less discussed other side of this conversation is that the role that changeling legends play in child abuse. There are numerous recorded historical instances of children being terribly abused or killed – Patrick's case is just one among many. Though filicide may seem incomprehensible, the staunchly held belief that the changeling was no longer a person removed traditional familial inhibitions and allowed for a great deal of atrocities to be committed. In fact, the concept of the changeling allowed parents to channel their emotions – frustration, grief, feelings of despair or failure – onto their child, since, of course, they viewed the person before them not as their child, but as a fairy substitute. The changeling legend thus adds a dangerous element to this already emotionally fraught experience by providing the rationale that families must "test" the changeling, and if it dies, they have simply done their job by ridding themselves of the fairy. This belief absolves the changeling's killers of any guilt – after all, even if the fairy substitute died, the human would still be living in the realm of the fairies. Patrick Doyle's death followed the pattern of many changeling tests – he died after being hit on the head with an iron hatchet three times (*Kerry Evening Post*, 1 Feb. 1888: 2). This action was likely rooted in several religious or traditional rationales. First, violence involving fire or iron was a staple of changeling tests, attributed to fairies' reported hatred for both fire and iron. Second, the ritualistic emphasis on the three strikes administered to Patrick may have had some religious or superstitious explanation – the number three is significant in both Catholic and Christian religions, often referencing the Holy Trinity.

When Patrick's body was discovered, his entire family, save his three young brothers, was arrested. Yet, only his mother, Johanna, remained in prison, serving a sentence for Patrick's murder. Though the entire family admitted to suffering from madness, some sort of group hysteria, Johanna was the only one to admit to striking Patrick. Was she a murderer, was she simply a mother trying to save her son in any way she could, or was she mentally unable to recognise that the boy before her was actually her child? The courts called her a murderer, but a common treatment for epilepsy was skull trephination. Perhaps Johanna was merely seeking to save her son's life by the only means available to her.

Modern portrayals of fairy changelings, such as *The Hole in the Ground* (2019), however, have recast the changeling children as an actual embodiment of evil whom parents must destroy in order to save their real child's life. As depictions of fairies and the fae evolve, popular focus seems to shift to a complete adoption of the horrific. No longer are changelings a fairytale concocted to rationalise real-life events; instead, they have become a tangible, magical, and malicious force in and of themselves. But despite the evolution of changelings, one thing remains constant: there is a deep-seated psychological component to the fear of child changelings that challenges parent–child relationships. This, perhaps, is what makes the idea of changelings so chilling, even today.

Changelings are undoubtedly moving from the historical to the horror, but they continue to offer parents – even in the 21st century – a way to grapple with the idea that ultimately, parents have no control over their "changing" children. Despite this lack of control of behaviour, personality, appearance, and even ability, parents have a societal expectation and obligation to care for their child and protect their best interests: even if this makes life more difficult. As in the case of *The Hole in the Ground*, Chris, the captured human child, has been making his mother's life increasingly difficult through social reticence, emotional outbursts, and struggles with his parents' separation. Chris is then replaced by a seemingly perfect changeling. Though the child was outwardly easier to parent, Sarah, the mother, immediately recognises that this is not her son and descends into paranoia and obsession with discovering the truth. But in a modern-day twist, unlike Patrick's mother, Johanna, Sarah does not kill the changeling until she has her real son safe in her arms. Perhaps there is a positive outlook on the societal role of changeling legends: rather than offering a real-life rationalisation of filicide, they now provide a hopeful perspective that children should be accepted as they are – even after experiencing "changing pains".

Saga Bokne

"Don't Call Me a Fairy" (Various, 1964–2006)

In Tad Williams' modern fairy story, *The War of the Flowers* (2003), the sprite Applecore reacts with violent indignation when jokingly addressed as Tinker Bell: "Call me that name again and you'll be wondering how your bollocks wound up lodged in your windpipe – from below. [...] 'If you believe in fairies, clap your hands!' If you believe in fairies, kiss my rosy pink arse is more like it" (262). Here, Applecore demonstrates her awareness of cultural history while simultaneously and decisively rejecting any suggestion of kinship between herself and J. M. Barrie's world-famous fairy. The fierceness of her reaction is all the more remarkable since the depiction of Applecore herself – winged, half a foot tall, wearing "a deep red minidress" (102), and possessing a large vocabulary of expletives – clearly owes a great deal to Peter Pan's legendary companion.

The interchange between Applecore and her friend is an example of what Michael Dylan Foster calls folkloresque parody or metacommentary. The *folkloresque* is Foster's proposed term for cultural works which are not in themselves folklore, but which allude to, appropriate elements from, or imitate folklore, often freely mixing folkloric materials from different traditions with entirely new creations, so as to achieve a fuzzy, difficult-to-pin-down "*sense of folklore*" (Foster 2016: 4, italics in original). Folkloresque *parody* or

---

1   Foster primarily uses *parody*, though he suggests *metacommentary* as an alternative term (2016: 18). In this chapter, I will use *metacommentary* which, I believe, better conveys the range of different functions performed by such self-conscious commentary.

metacommentary[1] is a subset of the folkloresque, characterised by a "seemingly intentional appropriation of folkloric motifs and structures for the purpose of caricature or similar modes of critical commentary" (18). It typically displays an "awareness of its own derivativeness" and is often explicitly self-referential (18). As such, it is part of a larger phenomenon of metareferentiality in the arts and media, a phenomenon which, according to some scholars (e.g. Wolf 2011), has played a role of unprecedented importance in cultural and artistic expressions of the last few decades. Folkloresque metacommentary can be used to comment on various targets, including "the source material (that is, folklore)", "itself as a popular culture product", or "popular cultural uses of folklore (that is the folkloresque)" (Foster 2016: 18). In this chapter, I will discuss selected examples of folkloresque metacommentary in a number of fictional works about fairies – we could call them meta-fairy-fictions, that is, *representations of fairies that comment on or refer to other representations of fairies*. Furthermore, I will suggest three major functions of such metacommentary in these fictions: *comedy*, *positioning*, and *undermining*.

The first function of folkloresque metacommentary in fairy fictions which I will consider is that of *comedy*. An example of a text where folklore itself is the butt of the joke is Lloyd Alexander's YA fantasy classic *The Book of Three* (1964). Captured by the Tylwyth Teg – the Welsh fairy people – the protagonists are brought before their king, "a dwarfish figure with a bristling yellow beard" by the name of Eiddileg (Alexander 2007: 136), and subjected to a rant on human behaviour:

> "If any of you thick-skulled oafs come on one of the Fair Folk above ground, what happens? You seize him! You grab him with your great hammy hands and try to make him lead you to buried treasure. Or you squeeze him until you get three wishes out of him – not satisfied with one, oh no, but *three*!"
>
> "Well, I don't mind telling you this," Eiddileg went on, his face turning redder by the moment, "I've put an end to all this wish-granting and treasure-scavenging. No more! Absolutely not! I'm surprised you didn't ruin us long ago!" (138–9)

The idea of the fairy king as a short, squat, red-faced figure, working himself up into a temper tantrum, is comical in itself because of the ways in which it clashes with received expectations of what a fairy king should be like. Moreover, in his accusations, Eiddileg refers directly to a folkloric context outside of his own fictional world. The behaviours that Eiddileg mentions do not occur in the novel itself; none of the characters in *The Book of Three* are guilty of having tried to squeeze a fairy for gifts or treasure. Instead, the humour hinges centrally on the readers' familiarity with previous representations of fairies – in this case, with the fairy tale genre, in which fairies are indeed known to grant gifts or lead humans to treasure. The implication of Eiddileg's rant is that the sheer number of such tales in the real world is a direct indication of the frequency of such occurrences in the fictive world – making Eiddileg's anger and frustration proportionate as, the reader must assume, his subjects can hardly venture outside their underground kingdom before a human seizes them and demands to have his or her wishes fulfilled. The funniness of the scene, then, is only accessible to a reader who catches on to this implication, and the recognition comes with the additional satisfaction of being part of a community, sharing cultural knowledge with the author.

In Eoin Colfer's YA fantasy series about the 12-year-old criminal mastermind Artemis Fowl (2001–12), folkloresque metacommentary is also employed for comic effect. Here, however, the folklore itself is not the major target of the parody; instead, the folkloresque is primarily used as a tool to poke fun at action movies of the gadget-happy kind, such as the *James Bond* or *Mission: Impossible* franchises. Nevertheless, as Foster notes, folkloresque metacommentary relies on an audience with knowledge both of "the popular culture product being critiqued" and "the folkloric elements invoked to enact the parody" (2016: 18). In *Artemis Fowl*, as in *The Book of Three*, the fairies are comical because of their perceived incongruity; we do not expect fairies to wield nuclear-powered guns, intercept human surveillance satellites, or mimic the hard-boiled jargon of cop movies. Apart from these implicit (though glaring) subversions of fairy conventions, *Artemis Fowl* also engages in explicit metacommentary. As in Alexander's novel, one of its chosen targets is the notion of fairies as conferrers of treasure. Artemis, having kidnapped the fairy police officer Holly Short, informs her that the goal of the venture is "to successfully separate a fairy from its gold". Her reply is deeply scornful: "Gold? Gold? Human idiot. You don't honestly believe that crock-of-gold nonsense. Some things aren't true, you know" (Colfer 2001: 120). Holly's retort is an example of a common rhetorical gesture in meta-fairy-fictions: the assertion that well-known tropes or beliefs about fairies are mistaken or untrue. Usually, readers are simultaneously provided with an explanation of how the purported misconception has come about. In this case, Artemis speedily assures Holly that, although he indeed used to believe in "all that under-the-rainbow crock-of-gold blarney," he is now aware of the true situation, namely the existence of a fairy "hostage fund" (121). A similar rhetorical move can be seen at work in the narrator's calm assertion that "the word 'leprechaun' actually originate[s] from LEPrecon, an elite branch of the Lower Elements Police" (33). Whereas, in *Artemis Fowl*, the claims of truth-telling are clearly tongue-in-cheek, very similar gestures are used in other fairy fictions with a much more serious air, as we shall see.

Turning to the next function of metacommentary in fairy fictions, let us go back briefly to the passage from Williams' *The War of the Flowers* discussed at the beginning of this chapter. The function of comedy is definitely relevant here; the sprite Applecore's caustic remodelling of the most famous line from *Peter Pan* is certainly funny. However, there is something else at work, too. Apart from making fun, Williams' novel, through Applecore, also engages in an act of *positioning*. Tinker Bell, especially as she appears in Disney's animated movie adaptation of Barrie's story, is likely the most commercially successful fairy of all time. She is the chief representative of a certain kind of fairy which, having been appropriated and mass-distributed by global culture industries, can be said to have attained the status of "global folklore" (see Peterson 2007: 94), fundamentally influencing the conception of fairies all over the world. By invoking Tinker Bell only in order to violently reject her, then, *The War of the Flowers* signals that this is going to be quite a different kind of fairy story – grittier, less sentimental perhaps, more adult. The commercialisation of folklore may also conceivably be part of the target.

Whether or not we are currently experiencing a "metareferential turn" as Wolf argues (2011: 1), this use of folkloresque metacommentary is not in itself new. Indeed, an illustrative example of such positioning can be found in Rudyard Kipling's *Puck of Pook's Hill*, originally published in 1906. Puck, "the oldest Old Thing in England" (Kipling 1994: 11), explains to the human protagonists why he does not like to use the word "fairy" to describe himself or his kind:

> [W]hat you call *them* are made-up things the People of the Hills have never heard of – little buzzflies with butterfly wings and gauze petticoats, and shiny stars in their hair, and a wand like a schoolteacher's cane for punishing bad boys and rewarding good ones. [...]

Can you wonder that the People of the Hills don't care to be confused with that painty-winged, wand-waving, sugar-and-shake-your-head set of impostors? (16)

Puck (and Kipling) here decisively distances himself from a type of fairy representation which was incredibly popular in Victorian and Edwardian England – and, it should be added, remains largely dominant to this day, Tinker Bell herself being an example of the type. Here, the truth-claim rhetoric can again be seen at work: the prettified, frail fairies of Victorian art and literature are branded as "made-up things" and "impostors". Against these allegedly false fairy representations, Puck posits a very different image: "Butterfly wings, indeed! I've seen Sir Huon and a troop of his people setting out from Tintagel Castle for Hy-Brasil in the teeth of a sou'westerly gale, with the spray flying all over the Castle, and the Horses of the Hills wild with fright. [...] Butterfly-wings! It was Magic – Magic as black as Merlin could make it" (16–17). Rejecting the fairies of Victorian nurseries and pantomimes, *Puck of Pook's Hill* instead aligns itself with a fairy tradition perceived as older and more authentic, where fairies are not frail, but ferocious; not sweet, but wielders of black magic. Doing this, Kipling's novel also appears to be set on a rescue mission, attempting to preserve the fairy as a legitimate subject for literature – even if, as Puck seems to suggest, the word "fairy" itself may be too far gone to be salvageable.

The narrator-protagonists of Keith Donohue's *The Stolen Child* (2007), a novel very different from *Puck of Pook's Hill* in most respects, nevertheless seem to be of an opinion similar to Puck's. "Don't call me a fairy", the changeling Henry Day begs his audience in the novel's opening line. "We don't like to be called fairies anymore. Once upon a time, *fairy* was a perfectly acceptable catchall for a variety of creatures, but now it has taken on too many associations" (Donohue 2007: 1). Exactly *which* associations of the word it is that are objectionable to Henry is not articulated; however, considering the generally bleak mood of the novel and the down-to-earth nature of its supernatural beings, it is feasible to assume that Henry, too, wants to distance himself from connotations of prettiness, childishness, and carefree whimsy. The use of "fairy" as a homophobic slur might perhaps also contribute to his scepticism towards the term.[2] After providing a brief lecture on the etymology of the word "fairy"

---

2   According to the *OED*, this usage is known at least since the late 19th century ("fairy").

and the categorisation of different types of spiritual beings, Henry suggests an alternative term for himself: "If you must give me a name, call me hobgoblin" (1). Both the lecture and the choice of a term with less literary and commercial baggage serve to frame Henry as an expert in the field, lending authority to his voice, and thus functioning as an implicit "truth claim" for his later assertions regarding the practices of faeries (as they are indeed called for most of the novel). This claim is corroborated by the second narrator-protagonist, the abducted child Aniday, who introduces *his* part of the narrative by asserting: "This is not a fairy tale, but the true history of my double life" (11). "Fairy tale" is here rhetorically opposed to "true history"; Aniday does not want to be associated with the flimsy fancies and bowdlerised myths of nursery literature. Instead, Aniday's claim and Henry's academic-style introduction both serve to position the novel as an adult, realist approach to the subject of fairies (despite the novel's clearly supernatural contents).

The third function of folkloresque metacommentary in fairy fictions, which I will discuss here, is rather more difficult to pin down. To outline this function, which I provisionally term *undermining*, I will use Michael Swanwick's extraordinarily complex and multi-layered science-fantasy novel *The Iron Dragon's Daughter* (1994) as my example text.

While Swanwick's novel does not offer any explicit commentary on the conventions of fairy representation, *The Iron Dragon's Daughter* is absolutely bursting with allusions to folklore, myth, mysticism, ancient religion, and literary works. The protagonist Jane – a changeling in a dystopian modernised Faërie – goes shoplifting in shops with names such as "The Eildon Tree" (Swanwick 2007: 76) and "House of Oberon" (128); rock bands with names like "Green Man," "Conjunction of Opposites," and "Wild Hunt" record albums such as "*Whitsuntide*" or "*Mythago*" (71, 83, 125); and the iron dragon of the title – a sentient fighter jet – harbours fond memories of atrocities committed in Lyonesse (41) and Avalon (98). Well-known fairy lore is appropriated, but with a difference: Fairy time reigns in the shopping mall, where shoppers can spend hours or days, yet emerge at the same moment at which they entered, and the fairies' Teind (tithe) comes in the shape of a one-night hedonistic riot, in which exactly 10 per cent of the population die, seemingly arbitrarily. The way folkloresque metacommentary is used in *The Iron Dragon's Daughter* is in many ways reminiscent of the comical uses in, for example, *Artemis Fowl* or *The War of the Flowers*. However, set in a grotesque, nightmarish world characterised by gratuitous violence and casual abuse, where moral values are unheard of and true friendship is extremely scarce, *The Iron Dragon's Daughter* does not primarily come across as comical. Nor does it attempt to align itself with any particular tradition or make claims of authenticity. On the contrary, the profusion of allusions, indiscriminately gathered from disparate sources, seems to deny and mock the very notion of an authentic tradition.

Tom Shippey argues that Swanwick's "heavy use of folklore motifs" provides a familiar ground which helps readers navigate the novel's otherwise unpredictable secondary world (2019: 427–8). I would argue, however, that the eclectic mix of folkloric and mythological references, together with literary allusions, snatches of pseudo-scientific discourse, and commercial slogans, constitutes a literary collage which rather has the effect of undermining the coherence of the storyworld. Patricia Waugh, in her seminal study of metafiction, writes that "[m]etafictional novels tend to be constructed on the principle of a fundamental and sustained opposition: the construction of a fictional illusion (as in traditional realism) and the laying bare of that illusion" (2001: 6). *The Iron Dragon's Daughter* is written within, and in response to, the fantasy and science fiction genres rather than "traditional realism". Yet, it still exhibits a "fundamental and sustained opposition". A basic generic characteristic of fantasy as well as science fiction is the creation of coherent secondary worlds – what J. R. R. Tolkien calls "sub-creation" (2014: 59). In *The Iron Dragon's Daughter*, Swanwick uses folkloresque metacommentary to simultaneously create and undermine his version of Faërie.

The cultural history of the fairy is long and variegated. Through the ages, fairies have appeared in many shapes and sizes and served many purposes, in traditional belief as well as in artistic expression. The writer or artist of today who wishes to engage with fairies has no choice but to negotiate this history of previous representations. As this chapter has shown, many writers choose to display, rather than hide away, this act of negotiation, demonstrating their knowledge of the fairies' cultural history through various uses of folkloresque metacommentary. Such knowledge may be revealed in order to be playfully twisted or parodied for the benefit of a knowledgeable audience, creating comedy. Alternatively, it may be used to position oneself, aligning one's creation with certain strands of fairy representation, while rejecting others. Or, thirdly, the history of previous representations may be embraced in its entirety, a multitude of conflicting images used in conjunction in order to undermine notions of authenticity and the possibility of coherent narrative. In all cases, however, folkloresque metacommentary serves to reveal, and educate readers about, the rich and varied cultural heritage of the fairy figure, while at the same time it enables writers to create something quite new out of old materials.

Francesca Bihet

A.I. Artificial Intelligence (Steven Spielberg, 2001)

The Programming of Fairy Belief in Steven Spielberg's A.I.

"Those were the years after the ice caps had melted because of the greenhouse gases and the oceans had risen to drown so many cities along all the shorelines of the world". With these opening lines, the narrator of *A.I. Artificial Intelligence* by Steven Spielberg (2001) inverts the normal "once upon a time" fairy-tale beginning, preparing viewers for a dark, unsettling, fairy tale of the future. Cringuta Pelea (2022: 5) points out that this narration generates "an aura of fairy tale and myth", but instead "depicts a dark and nihilistic world, a futuristic dystopia in which the human species has destroyed itself through global warming". In this genre-bending film, fairy-tale mixes with science fiction, posing questions about the nature of humanity, and its relationship with the natural and supernatural. It is a twisted fairy tale, with features of *Pinocchio*, that become skewed and disturbing. Set in the United States after the deluge, where poorer parts of the world starve, while the privileged human "orga" still enjoy a high living standard maintained by strict birth control and the services of robotic "mecha". Frances Flannery-Dailey (2003: 2) describes the film as "an illustration of intelligent, postmodern myth-making that constructs a multi-layered reality by interweaving dreaming, technology, ontological confusion, non-linear time, religion and myth". The film, which divided critics, poses pertinent questions for our contemporary world, with AI chatbot technology introduced into our daily lives and the increasing urgency of climate change. Like the fairy-tale genre, Isabella Hermann (2023: 319) argues that the science fiction "serves as a distorting mirror and metaphor to reflect on the human condition and socio-political issues in relation to and beyond technology". This chapter examines how fairy tale and *Pinocchio*'s Blue Fairy is used symbolically throughout the film to explore human nature and our belief systems. By confronting audiences with a distorted reflection of our own society, the film helps us examine what fairy tales can tell us about our own beliefs and culture. Fairies and fairy tales express our deepest human fears and hopes, likewise this film taps into a deep vein of fairy-lore, showing that belief in the supernatural is deeply engrained in human culture, even in a world filled with high technology. The fairies in *A.I.* might manifest as holographs, but they represent a traditional storytelling element, which has endured with humanity since the beginning.

Based upon Carlo Collodi's fairy tale *Pinocchio* (1883) and Brian Aldiss's "Super-Toys Last All Summer Long", first published in *Harper's Bazaar* in 1969, Professor Hobby programmes robot boy David as an image of his own dead son, a loving child living within an idealised American family in an over-populated future. This surrogate robotic child is given as a trial to Monica and Henry, a Cybertronics employee, whose critically ill son Martin is kept in cryogenic suspension. When Martin recovers and family relations become strained, Monica abandons David like a changeling into the cruel world, vulnerable to the robot destroying "Flesh Fairs". Naarah Sawers (2010: 48) notes that this theme, like "Hansel and Gretel" or "Babes in the Woods", "amounts to an aggregate of fairy tale references that pre-empt childish adventures". Inspired by unconditional love for Monica, his mommy, and deep belief in the fairy tale *Pinocchio*, David goes on a quest to find the Blue Fairy, who will make him a real boy. His wish is only ostensibly fulfilled when he gets re-found by some alien-like mecha 2,000 years later, trapped in an amphibicopter in the sunken Coney Island amusement park near a Blue Fairy statue, after all humanity has perished. Like Pinocchio, David seeks validation and love as a human child and not a machine. James Naremore (2005: 8) highlights the "lineal relationship between Collodi, Disney, and Spielberg". Many storylines in the film reference these predecessors. Flannery-Dailey (2003: 22) suggests numerous parallels between *A.I.*'s narrative and Disney's *Pinocchio*, including "the guide and companion Jiminy Cricket/Teddy", "Geppetto/Dr. Hobby", "imprisonment in Stromboli's Marionette Circus/Flesh Fair", both characters' plunge into the sea to find their parental figures, and most obviously the Blue Fairy. Spielberg's dark version is reflective of 19th-century children's fairy tales, where stories often had more stark moral lessons, than the sanitised 20th-century Disney films. For example, in Collodi (1916: 25–7) the poor cricket is killed by Pinocchio when trying to deliver a moral message. The film is fully conscious of the fairy-tale landscape it has inherited, referencing the darker elements from Collodi, as inherited through the legacy of Disney's film. *A.I.* is the next retelling, the future of Pinocchio.

The Blue Fairy in *A.I.* manifests as an unattainable remote goddess figure, who David is always striving to reach; quite unlike the Fairy with the Blue Hair in Collodi's *Pinocchio*. In *A.I.* the fairy is an un-present presence; her symbolism incorporates multiple aspects of both the divine and mortal feminine. Thomas Morrissey (2004) highlights the difference between the folk-fairy in Collodi "a crafty shape-changer" and "the Disneyesque icon" upon which David fixates. When Monica reads Martin and David *Pinocchio*, David is primed for his adventure, and given faith that the Blue Fairy is a loving maternal fairy godmother, unlike Monica. While the Blue Fairy is not present as a character in the film, David's faith in her drives his adventure. Collodi's fairy is the archetype 19th-century fairy godmother from children's literature, a wish granter, protector, and ever-caring moral guide. Sabina Magliocco (2019: 115) highlights that fairy godmothers are "a child's idealised, fantasy version of adults", they "help young people in the process of maturation and the attainment of social status". Although, in some examples, such as Charles Perrault's "Toads and Diamonds" included in Andrew Lang's *Blue Fairy Book* (1889: 274–7), these figures do scold children to help develop moral improvement. Collodi's fairy is no different. She provides moral lessons, like the famous enchanted nose which grows when Pinocchio lies and teaching Pinocchio to take his medicine by frightening him with rabbit pull-bearers (Collodi 1916: 83–9). As Morrissey (2004) notes she helps Pinocchio "accentuate his finer character traits by means of some pretty nasty tough love". However, in *A.I.* David focusses upon the wish-granting nature of the Blue Fairy. Like Cinderella's fairy godmother, Collodi's Blue Fairy rescues Pinocchio in an enchanted coach, conjured with two claps, "drawn by a hundred pairs of white mice", with a poodle coachman (Collodi 1916: 79–80). However, Morrissey (2004) points out that whilst Pinocchio experiences moral improvement and growth into an "industrious and caring puppet", one who can "mature", David has "no-growth programming", he will remain a child. As Collodi's *Pinocchio* (1916: 133) states, he wants to become real because "puppets never grow. They are born puppets, live puppets, and die puppets". Morrissey (2004) notes that this represents the "ultimate tragic core" of David's character. David is denied real growth, the fairy doesn't actually help him in the film, she is a figure upon which he is transfixed to resolve his unfulfilled desires. A boy that will never grow up, David will continue to believe in fairies forever.

Symbols of the Blue Fairy abound throughout the film, such as the virtual image in Dr Know's knowledge vending machine, the Coney Island statue-turned-icon, and as a holograph conjured by the futuristic mecha who eventually find David (Tibbetts 2001: 258). Yet the Blue Fairy never actually manifests. She represents multiple aspects of femininity beyond the fairy godmother archetype. David's companion, Gigolo Joe, an escaped lover mecha, asks about the nature of the Blue Fairy. He states, "In the world of orga, blue is the colour of melancholy. The services I provide will put a blush back on anyone's cheek. I will change the colour of your fairy for you. She will scream out in the moonlight". Joe sees the fairy as a potential client, he merely desires to perform his programmed function as a lover robot. The Blue Fairy here appears as a sexualised female. Likewise, Joe guides David to Rouge city, a futuristic city of pleasure where buildings are shapely female figures. They ask Dr Know about the Blue Fairy under the category of fact, and it provides information about the Blue Fairy escort agency. When Dr Know shows a projected image of Collodi's fairy, David chases after the hologram, grasping at the air. She remains repeatedly symbolic, "an example" as Joe notes, unobtainable, forever just out of reach. Indeed, much of the later section of the film is bathed in blue-hued lighting; the Blue Fairy becomes a misty presence over everything that never fully materialises. Hobby portrays himself as playing the function of the Blue Fairy, as David's creator. He states, "But you are a real boy. At least as real as I've ever made one. Which by reasonable accounts would make me your blue fairy". Hobby is cold and logical, disclosing the horrible truth that the Blue Fairy does not exist. This mirrors the death of the Blue Fairy in Collodi's Pinocchio (1916: 116), when he finds her tombstone, declaring the loss of Pinocchio as her cause of death. These scenes become emotional turning points for the character, when all hope besides belief is lost. When David discovers he is a consumer product and presses his face inside the mask of one of the unfinished robots, surrounded by blue lighting, Carolyn Jess-Cooke (2009: 146) suggests that this "configures David himself as the Blue Fairy". Throughout the film David is searching for the Blue Fairy, but in the end, she is a matter of faith, found only within hm.

David becomes fixated upon the divine feminine aspect of the Blue Fairy, eventually worshiping her like a goddess. It is this aspect of her which is most powerfully portrayed throughout the movie. In Rouge City, David asks Joe if a figure of "Our Lady of the Immaculate Heart" outside a chapel is the Blue Fairy. Joe replies: "The ones who made us are always looking for the ones who made them. Go in, fold their hands, look around their feet, sing songs, and when they come out it is usually me, they find". Joe's mechanical observation on human religious practice is a bleak reading of religious faith. Nevertheless, as Flannery-Dailey (2003: 14) points out, the film generates a symbolic and iconographic parallel between the Blue Fairy and the Virgin Mary, as a life-giving mother. Indeed, just before David dives into the all-consuming blue ocean, he simply says "mommy", a feminine inversion of the scene in Collodi's *Pinocchio* (1916: 122) when he dives into the water in an act of love declaring: "I will save my papa!". In what V. Alan White (2008: 210–26) terms the "Blue Fairy problem", he argues that the scene demonstrates David's belief revision doubting the Blue Fairy, unimaginably glistening below the blue water, and acts with momentary despair, which is a sign of true human reasoning. Yet contrary to this previous evidence, when a model of the Blue Fairy appears like an icon as part of the submerged Coney Island *Pinocchio* display and David's amphibicopter gets trapped, he prays incessantly to her for 2,000 years, as humans pray in the Chapel of "Our Lady of the Immaculate Heart". The narrator tells viewers, "[a]nd David continued to pray to the blue fairy before him. And she who smiled softly forever, who welcomed forever". A "blue ghost in ice", she is "always there, always smiling, always awaiting him". Seemingly, only humans (and David) have the capacity to bestow faith in the unseen supernatural, which always lies just beyond the tangible. Tim Kreider (2002: 37) notes the "serene face", conjoins Monica and the Blue Fairy in one, "an image of emotional fusion" for David. The moment David is freed by the advance mecha, he goes over to touch the Blue Fairy statue, which shatters in the ice. All his dreams and hopes crumble before his eyes, once again revealing the truth.

    David's faith in the Blue Fairy, his belief in the fairy tale, demonstrates his humanity beyond the capacity of other robots. Hobby describes him as "[a] mecha with a mind", with "a kind of subconscious never before achieved, an inner world of metaphor and intuition, self-motivated reasoning, or dreams". Hobby is amazed when David is drawn back to the Cybertronics headquarters after following the fairy-tale riddle from Dr Know's booth, citing W. B. Yeats' poem *The Stolen Child* (1908: 39–42) about a child captured by the fairies. This riddle sits liminally in Dr Know's taxonomy, created by merging the "flat fact" and "fairy tale" categories. Hobby states, "[u]ntil you were born robots didn't dream". Hobby describes the Blue Fairy as "part of the great human flaw, to wish for things that don't exist", "the ability to chase down our dreams". Hobby sees David as a programmed human, displaying faith and belief. Joe had previously stated that only "orga" could believe in the supernatural.

> What if the blue fairy isn't real at all David? What if she is magic? The supernatural is the hidden web that unites the universe. Only Orga believe what cannot be seen or measured. It is that oddness that separates our species. What if the Blue Fairy is an electronic parasite that has arisen to haunt the minds of Artificial Intelligence.

Joe is scared that the Blue Fairy might be some kind of programming error or virus, causing David to repetitively fixate on the Blue Fairy. This commentary hints that belief in the supernatural might also be a glitch that is unique to humans. Daniel Dinello (2005: 84–5) sees David's belief in the Blue Fairy as a "self-destructive programming glitch". Yet parallels with religion in the film, show that belief is, in fact, a human feature of the robot boy, to hope and dream without reasonable justification, when all appears lost. Like humans, David never gives up hope in the Blue Fairy, despite moments of lapse in faith, he always returns. Kreider (2002: 34) points out that David's searching and suffering is "irrational, unconscious – what we might call hardwired", it also "makes him a tragic figure", and in ways "his manufacturers never intended, also what makes him human". David's "obsession" (Jackson 2017: 59) with the Blue Fairy can either be viewed as an electronic parasite, a programmed glitch which he cannot question (Heffernan 2018: 14), or indeed, that which displays his human-like capacity to hold unerring beliefs. Perhaps, Spielberg is highlighting that belief in the face of all doubt might be a human flaw, but it is also a beacon of hope when all seems lost.

Ironically in the last scenes, a holograph of the Blue Fairy conjured by the futuristic mecha does help David, finally granting his wish of acceptance by his mother. These future mechas value David as a unique being, as Dinello (2005: 85) notes, he is "studied as the link to an extinct human species that the future humanoids strangely admire". His memory chip, with data about

A.I. Artificial Intelligence (Steven Spielberg, 2001)

the now destroyed human world, becomes an archaeological artefact hidden in the ice. As Naremore (2005: 12) points out, the mechas look down upon David's memory, "as if they were archivists looking down at an old movie that offers a key to the human psyche". These mechas provide David with "fulfilment", Flannery-Dailey (2003: 13) points out, using his memories "they simulate the Blue Fairy" to "reconstruct his ideal reality". The Blue Fairy holograph tells David in an echoey, dream-like voice: "You are unique in all the world". Throughout the film, which is framed as a fairy-tale quest, the Blue Fairy is notably missing as a character to support and aid David. Finally, the fairy who was always out of grasp, yet calling, seemingly materialises. The alien-like mechas perform the functional role of the Blue Fairy, which is fitting considering, as Diane Purkiss (2000: 3) notes, that in the modern world "Aliens are our fairies, they behave just like the fairies of our ancestors". A note of hope is found in these "kind and benevolent" mecha, who have seemingly lost humanity's flaws (Jackson 2017: 60). These mechas give David one wonderful day with Monica, which ends with them falling asleep forever, as the narrator states David returns "to that place where dreams are born". The fairy-tale ending is deeply sentimental. However, the fulfilment of his wish, a form of conservation or reconstruction, also confers recognition of David's unique status. Holly Blackford (2007: 78) points out that folk characters are "insignificant when thrown into a world they do not control", but finally throughout the narrative "they need to become equal to their creators". The conjured Blue Fairy acts as a pastiche fairy godmother, making David the protagonist of his own fairy tale, not a replicated production-line robot. Kreider (2002: 37) points out that the mechas tell David, "He *is* one of a kind, treasured now as a singular, irreplaceable artifact, that he is loved". Finally, David is acknowledged and the fairy tale is complete. The fairy-tale mode, as Heffernan (2018: 11) notes, "clashes sharply" with the wider science-fiction apocalyptic framing. Yet Joshua Sikora (2020: 272) highlights that devices, such as narration, lead to the "optimistic arc of fantasy", which "asks us to submerge ourselves into the dream, accepting the abstraction and projections with the faith of a child". The film presents us with a false, simulated, fairy-tale ending. Whilst viewers remain all too aware of humanity's destruction, David's hope remains.

Part II
Belonging and Otherness

Angana Moitra

Queene of Phayries (Various, 1590–1602)

Fairy Mythology as Instruments of Genealogy Embellishment

The use of fairy mythology in the literary culture of the British Isles has had a chequered history of representation, vacillating from the sinister otherworldliness of *Sir Orfeo* to the petulant and comic parallelism of the fairy kingdom (headed, interestingly enough, by the pagan deities Pluto and Proserpina) in the "Merchant's Tale" of Geoffrey Chaucer's *Canterbury Tales*. Such oppositional attitudes represented not only the capacious polysemy of medieval conceptions of the supernatural, but also the perennial polarisation between the folkloric roots of fairylore and the theological anxiety generated by the Church's attempts to explain the "pagan" provenance of fairies (Cooper 2004: 173–217; Green 2016; Saunders 2010: 179–206; Wade 2011; Watkins 2007). Indeed, perhaps more than any other subset of the supernatural, the ontological fluidity and ambiguity of the fairies ensured that they were amenable to incorporation within a variety of (often conflicting) registers. However, by the 16th century in England, one of the primary modes of representation of fairies was as the supernatural ancestors of ruling families. As dynastic houses jostled for primacy and power in the fraught political landscape of early modern England, they sought to construct familial lineages which were frequently predicated upon fairy genealogy. A similar strategy was adopted by the Tudors who claimed descent from fairy forefathers (and, occasionally, fairy foremothers), a strategy of self-fashioning which peaked under the reign of Elizabeth I (Greenblatt 1980; Woodcock 2004). Such representational paradigms were most frequently deployed in the civic pageants and private receptions arranged to pay homage to the English queen.

The network of associations between political exigency, dynastic legitimation, and fairy ancestry can be traced back to the genre of the medieval romance, specifically the romances centred on the figures of Melusine and Arthur (Harf-Lancner 1984). The first version of the Melusine story had been composed by Jean d'Arras under the patronage of Jean, Duc de Berry, second son of King Jean II of France and the count of Poitou. Jean d'Arras' story, which related the story of how the fairy Melusine had founded the Lusignan line through her union with Raimondin, was intended to simultaneously vindicate the Duc de Berry's claims upon Poitou (which had been one of the nerve centres of French resistance to the English during the battle of Poitiers in 1356) and bolster French nationalist aspirations against competing English claims in the Hundred Years' War (Péporté 2017: 162–79). By figuring the Duc de Berry as the successor to a dynasty which was ostensibly birthed by a legendary fairy foundress, Jean d'Arras', fairy romance not only championed Poitou's political status but also strengthened French Angevin hegemony over the English crown. However, given the complicated reproductive politics of Anglo-French bloodlines in the Middle Ages, the ideological cachet of using Melusine's fairy status to claim political supremacy was attractive not only to the French but also to the English. As fairy ancestress, Melusine was cited as the progenitor of a number of European noble houses, including the family of Elizabeth Woodville, wife of England's Edward IV, and this association was exploited by the author of the *The Romans of Partenay, or of Lusignan*, a 16th-century Middle English translation of the Melusine legend based both on Jean d'Arras' foundational text as well as Coudrette's later verse text (Alberghini 2017: 146–61).

Insofar as the British Isles is concerned, however, the primary inspirational force for grafting fairy mythology to the imperatives of insular politics is provided by the figure of Arthur rather than Melusine. First appearing in the 12th century in Geoffrey of Monmouth's *Historia Regum Britanniae*, a compendious (albeit largely fictitious) digest of England's regnal past, Arthur proved to be irresistible to versifiers, prose writers, and chroniclers alike, reappearing almost incessantly in medieval textual culture (Putter 1994: 1–16; Rikhardsdottir 2017: 135–50). Just as Melusine could be cited as supernatural progenitor to buttress French nationalist ambitions, Arthur became the locus around whom ideas of political deliverance and imperial enfranchisement constellated (Summers 1997). The notion of Arthur-as-messiah was a particularly potent one for the fringe lands of Wales and Brittany who rallied behind his iconic legacy to oppose French territorial expansion (Berard 2019). The association between the corpus of medieval Arthuriana and fairylore was an early one, provided by the connective link of the figure of Morgan le Fay, variously depicted as evil adversary, benevolent healer, supernatural enchantress, and disputed familial relation (Enstone 2011; Hebert 2013). This interpellation of fairy mythology and insular historical record intensified through the course of the Middle Ages, and by the time Thomas Malory inherited the apparatus of the Arthurian legendarium in the 15th century, the matrix of associations between imperial ambition as embodied by Arthur and the supernatural elements of fairy magic (such as the figures of Morgan and the Lady of the Lake, the sword Excalibur, and the enchanted isle of Avalon) had been firmly established (Moitra 2023). The dynastic houses of western Europe were alert to the potential of the Arthurian cycle to be used as instruments of genealogy embellishment, and within the British Isles, Arthurian lore had been co-opted by monarchs and ruling families from Henry II onwards to legitimate the (political and moral) "rightness" of their rule (Aurell 2007: 362–94; Johanek 1987). This process, which began with the Plantagenets, was to have its apotheosis at the hands of the Tudors.

Although historical precedent provides a convenient point of reference to contextualise the Tudors' use of fairylore, the ideological reasons behind such a choice are more difficult to explain. In such a case, Antonio Gramsci's theory of hegemony can provide a template to decode the relationship between fairy mythology and Tudor performative culture. According to Gramsci, hegemony is the power exercised by a dominant group through the manufacture of consent for the purpose of maintenance of its supremacy as well as for the preservation of the status quo. Such consent is obtained through the provision of intellectual and moral leadership rather than the exercise of force or coercion (Adamson 1980; Femia 1981; Grelle 2017). Hegemonic control aims at the internal realignment of personal convictions (via consensual agreement) to the mass validation of prevailing norms and is achieved through the operation of educational, religious, civic, and political organisations which together comprise the fabric of society (Bates 1975: 353). One of the ways in which the dominant class ensures its hegemonic control over subordinate groups is through the appropriation and calibration of the devices of popular culture, a category which includes, among other things, the diffuse and rich world of folklore. According to Gramsci, folklore relates to a particular conception of the world of certain social strata untouched by modern currents of thought (Gramsci 1992). As a reflection of the conditions of existence of the people, folklore forms a distinct species of the popular culture of subordinate groups and is based upon the interpretations of quotidian existence practiced by the masses on the level of their own intellectual, moral, and religious level. Hegemonic powers attempt to elicit consensual support for their regimes of operation through an ideological contest with subordinate groups played upon the cultural field of folklore. According to Gramsci, dominant groups have their own conception of life which they strive to propagate among the masses. This dissemination does not, however, take place on a *tabula rasa*; rather, it is a competitive manoeuvre which often clashes with folklore and must overcome it. One of the ways in which such an ideological conflict is resolved is, paradoxically, through the dominant group's appropriation of folklore. As the folklore of subordinate groups is expropriated and remoulded by the dominant group, the continuance of hegemony is assured through the maintenance of consensual approval.

Applying the Gramscian formulation of hegemony and the attempt of hegemonic powers to ensure their control over subordinate groups through the appropriation of the popular culture of the masses (in particular, through the extrapolation and manipulation of folklore) to 16th-century England, a rationale can be found to explain the utilisation of fairy mythology by the Tudor regime in the pageants, displays, and processions which constituted the performative culture of English monarchy, especially during the reign of Elizabeth I. As the dominant political power (one which, despite its Welsh provenance, had not only wrested control of the English throne but had also reoriented the religious culture of the nation), the house of Tudor can be seen as a hegemonic power insofar as it managed to secure popular consensual support for its right to rule. One of the ways in which it generated such consensus was through the careful curation of the imagery and aesthetics of a representational culture of performance and textual depiction which relied heavily upon fairylore.

Fairies were utilised both during Elizabeth's civic progress through Norwich in the summer of 1578 as well as the private receptions at Kenilworth, Woodstock, Ditchley, and Elvetham hosted by select members of her inner circle. During the Norwich progress, fairies were included in the pageantry which concluded the five-day celebrations devised by Thomas Churchyard to regale the queen. In an interesting act of substitution (admitted by Churchyard himself), fairies were chosen to replace the entertainment by the water nymphs which had been planned for the previous day, but which had to be abandoned due to heavy rainfall (Nichols 1823: 211). Although Churchyard does not explain why he specifically chose *fairies* to replace water nymphs (practical reasons could be that dry land was preferred to an aquatic spectacle in the face of inclement weather and the costumes of the water nymphs could be easily repurposed for the fairies), the reasons behind his choice matter less than the selection. In the Norwich entertainment, Elizabeth was addressed by the "Queene of Phayries" who described how the fairies had been silently and faithfully keeping watch on Jove's orders, waiting to bid her farewell as she departed from the city (Nichols 1823: 212–13). The Fairy Queen's speech not only neatly links the entertainments which Elizabeth had witnessed over the course of the week but also establishes fairies as intermediate creatures, poised between classical gods and Christian spirits ("saints and soules, and sprites of men") and imbued by the ardour of love and the solemnity of duty.

Prior to Norwich, fairies had already appeared in the receptions hosted in 1575 by Robert Dudley, Earl of Leicester at Kenilworth, and by Sir Henry Lee at Woodstock. At Kenilworth, the figure of the Lady of the Lake together with her connections with Arthurian legend was used as part of the broader chivalric setting which framed the celebrations. The Lady's welcome speech greeted Elizabeth with a (fictitious) account of Kenilworth's august lineage of possession, and although the register of the speech was firmly within the tradition of claiming fairy ancestry that had become an established part of ideological praxis, its purpose in this case was probably to amplify Leicester's suitability as a potential husband for Elizabeth (Goldring 2007: 163–88). Over the course of the performance, Elizabeth is also transformed from passive addressee to active agent when she is requested by Tryton to free the Lady from her imprisonment by Sir Bruse sans pitie (Nichols 1823: 420–523). At Woodstock, Elizabeth is welcomed by the Fairy Queen in a speech which simultaneously extols the virtues of the English queen and characterises her as the fairy's close "frende". The Fairy Queen also acts as supernatural intercessor to resolve the political (and emotional) entanglements at the kingdom of Cambaya in a dramatic interlude devised by Lee (Cunliffe 1911: 92–141).

At the reception hosted by Edward Seymour, Earl of Hertford, at Elvetham in 1591, Elizabeth is once again greeted by the Fairy Queen (here named Aureola) with a garland shaped like an imperial crown and by a speech peppered with allusions to pastoral and pagan myth. Aureola's address also makes a reference to Auberon, the Fairy King, who is presented as the one who had sent his wife to deliver the royal welcome (Wilson 1980: 99–118). One of the final performative representations of fairies in the Elizabethan entertainments occurred at the Ditchley progress in 1592. Devised once again by Lee, this progress marked a continuation of elements presented 17 years earlier at Woodstock. The Fairy Queen makes an appearance again at Ditchley, although her presence is indirectly indicated by the verbal report of a knight guarding an enchanted grove. However, both the register and the frame of reference have changed: at Ditchley, the Fairy Queen is revealed to be the prisoner of souls, a "just revengefull" enchantress whose "infernall Arte" has incarcerated numerous knights and ladies whose transgressions – if any – are not revealed (Nichols 1823: 198–213; Wilson 1980: 126–42). It is not clear whether Elizabeth and the Fairy Queen are friends or rivals by this point, since although the knight refers to the Woodstock entertainment (where the two had greeted each other like equals), he also enjoins the English queen to undo the fairy's enchantment and liberate the prisoners of the grove. Such uncertainty is, however, not only in keeping with the ontological ambivalence of fairies, but also hints at the polarities – just and honourable but also malicious and unpredictable – inherent in the nature of Elizabeth herself.

By the beginning of the 17th century, fairy mythology in the British Isles had become a particularly fluid hermeneutic field, a core cultural bloc which could be mined for ideas, tropes, and motifs as the situation demanded. Given the historical association between fairies and the imperatives of imperial legitimation (as embodied in the legends of Arthur and Melusine), dynastic houses found in fairylore a potent metaphor to give credence to their political ambitions. The deployment of fairy symbolism by ruling families (especially the House of Tudor) can be read as an attempt to manufacture consent and mass approval by dominant groups as theorised by Antonio Gramsci. To ensure the longevity and power of such consensual relations, hegemonic authority relies not on coercion but on cultural constructs such as folklore which can be manipulated and remoulded according to the requirements of political expediency. A similar strategy was adopted by the English monarchy under the Tudors, especially during the reign of Elizabeth. By consistently utilising the semiotics of fairylore in the civic progresses and private receptions that accompanied the sojourns of her peripatetic court, Elizabeth I could use a visual and ontological apparatus which was at once delightful visual spectacle, powerful political sign, and instrument of genealogy fashioning. At the same time, the queen could also rely on the fairies' status as a subset of the ambiguous supernatural to claim for herself an ambivalence which qualified her reputation for justice, temperance, and moderation by reminding both civic audiences and courtiers of her capacity for ruthlessness and even violence.

Joan Ormrod

Jonathan Strange and Mr Norrell (Susanna Clark, 2004)

Fairies, Landscape, and National Identity

Jonathan Strange and Mr Norrell (Clarke 2004) is set in an alternate 19th-century England and tells the story of Gilbert Norrell (hereafter Norrell) who revives the fiancé, Emma Wintertowne, of cabinet minister, Sir Walter Pole, with the aid of a mysterious gentleman with thistle-down hair (hereafter, the Gentleman). He promises half of Emma's life to the Gentleman in payment, little realising that requires her to travel to the Gentleman's fairy mansion, Lost Hope, every night to dance until first light. Norrell's subsequent fame attracts a disciple, Jonathan Strange (hereafter Strange), who becomes a magician on a whim when given two spells by Vinculus, a street magician. However, the two soon clash on their approaches to magic, not least as Strange, unlike Norrell, believes that John Uskglass (hereafter Uskglass), the Raven King, who was raised as a slave in Faerie and ruled the North from 12th to the 15th centuries, before disappearing in 1485, can return British magic to its former glory. Much of the story is influenced by the Gentleman, a fairy servant of Uskglass; he encounters Strange's wife Arabella, who later seemingly sickens and dies but, like Emma, she is taken to Lost Hope; he enslaves Stephan Black, Sir Walter Pole's butler – who was given a prophecy by Vinculus that he "shall be a king in a strange country" – compelling him to take part in his wicked antics, such as hanging Vinculus. Meanwhile Strange manages to enter the fairy realm through madness where he discovers Lady Pole and Arabella in Lost Hope. Strange and Norrell restore the relationships that exist in England between the forces of nature, who were ruled by Uskglass, and bestow the power of English magic on Stephen Black, allowing him to destroy the Gentleman. Stephen then disappears from England to rule over Faerie. Vinculus is restored, the tattoos on his body are revealed as the last book of magic written by John Uskglass and magic returns to England.

This chapter uses examples from novel and television series to explore how fairies represent British national identity within the British landscape and its historical setting. Fairies within the English landscape represent chaos and the eruption of fantasy into a rational world. Fantasy has a tendency to erupt despite the British turn towards control and sensibility for, as Peter Ackroyd (2002) argues, from the earliest times, the English imagination is hardwired to natural phenomena such as water and landscape. Fairies are aligned with the elements of the British landscape, especially the remnants of our ancient past. According to John Kruse (2022) "faeries connect us directly and physically to Britain ... we are drawn to sites such as ancient prehistoric monuments precisely because they are symbols of this ... they are a tangible manifestation of a deeper psychic bond" (4). Accordingly, this chapter first explores and shows how national identity formed within the specific epoch and the importance of landscape in its construction. Ancient monuments were integral to this construction and fairies often inhabit the unexplained bumps and tombs in the landscape. The final section analyses the fairy in the story with landscape particularly in the association of death, resurrection, trees, and prehistoric sites.

The Emergence of the Modern Era in the Early 19th Century

Diane Purkiss (2000) notes that fairies emerge in times of transition or disruption and this story is set at a pivotal or unsettled moment in history after the French Revolution and in the Agrarian and Industrial Revolutions (Purkiss 2000). The Agrarian followed by the Industrial Revolution prompted the migration of people away from the countryside to the city. Society became more secular and the tensions produced from the move between rural and city developed debates between the Classical versus the Romantic, which was principally pronounced in the Gothic movement in the arts. This era, according to Michel Foucault in *The Order of Things* (1994), signals the beginning of the modern epoch when, "different kinds of change were taking place in scientific discourse – changes that did not occur at the same level, proceed at the same pace, or obey the same laws" (xii). It represented a shift from the Medieval mindset predicated on the supernatural and religion, towards empirical knowledge and science. Britain was the first country to embrace this change, but encounters with different cultures, flora and fauna in unknown parts of the world, prompted debates on national identities across Europe. Classifications of human versus others were reinforced by scientific analysis (Foucault 1995). The ideological conflicts between Strange and Norrell replicate similar debates circulating in the time in which the story is set: the late 18th and early 19th century. Such debates also found expression in the English landscape between the wild, primitive, and uncivilised places of faerie and the attempts of humanity to bring order and rationality.

Susannah Clarke's physical world is constructed through the concept of animism, the belief that the material universe is organised through a supernatural power, whether inanimate, living, or natural phenomena. Fred Ingold (2011) describes animism as "a sense of being in the world – heightened sensitivity and responsiveness, in perception and action to an environment that is always in flux" (9–10). Throughout *Jonathan Strange and Mr Norrell*, there are statements to support the underpinning notion of animism in this universe, for instance, in Norrell's first demonstration of his "practical magic" when he brings to life the stones of York Minster. Norrell also tells Strange that "Fairies do not make a strong distinction between the animate and the inanimate. They believe that stones, doors, trees, fire, clouds and so forth all have souls and desires" (466). The magic of the landscape enables Strange, Black, and Norrell to defeat the Gentleman and the fairy is also an integral part of the landscape.

British Landscape, Heritage, and Fairies

Landscapes have never been unchanging concepts but moulded through time, environment, and human intervention. John Agnew (2016) states that "Landscapes are ... material custodians of both historical memory and the sense of place" (1). According to David Lowenthal (1991), the British landscape, like nowhere else is fetishised for its beauty which differentiates it from landscapes anywhere else in the world. Lowenthal attributes this to British landscape's alignment with heritage rather than history. Heritage, like the fairy glamour, confers a superficial beauty on the landscape giving it symbolic and mystic power to the extent that, "[e]ven early megalithic civilization is lauded for adding a new dignity to Nature and leaving English country actually more beautiful than it found her" (Lowenthal 1991: 213). The earlier layers are always present in unexplained bumps and strange features often regarded as the houses of fairies (see Figure 6).

The British landscape is incorporated into a national myth-making process in which nostalgia for the heritage past is articulated through landscape. We fetishise the British landscape possibly because of our isolation at key

Figure 6. Fairies reputedly lived under prehistoric sites in folklore, for example, Bincombe Bumps, Dorset. Photograph by John Sayer for "The Cereologist" at www.sayer.abel.co.uk, and used with permission.

moments in history. The early 19th century is significant because it is a time when Britain was isolated through war (Hughes and Heholt 2018) and society disrupted with the Agrarian and Industrial Revolutions. It is this moment when fairies emerge in the story.

Debates about control over the material world as opposed to the unknown were played out in the development of romanticism near the end of the 18th century. Such debates centred on nature and the British landscape as either uncontrolled, wild, and sublime or contrived and picturesque (Hussey 2018). The picturesque phase is described as "the point when an art shifted its appeal from the reason to the imagination" (Hussey 2018: 4). The excesses of the imagination were reflected in the Gothic aesthetic (Botting 2014; Hughes and Heholt 2018). The Gothic movement which emphasised the mysterious, the supernatural, heightened emotions, the ugly and sublime, and excess, and can be regarded as artist's responses to the scientific and rational examinations of the natural world.

Fairies reflected this Gothic turn in aesthetics and they are often aligned with the heritage landscape. Like other magical beings such as the sleeping King Arthur and giants in British folklore, fairies are aligned with prehistoric sites in the landscape that connect the mortal realm with the underworld such as burial mounds, henges, and brughs (Briggs 1957, 1967, 1970; Houlbrook and Young 2018; Kruse 2022; Sugg 2018;) (see Figure 2). David MacRitchie (1890) argued that fairies were part of prehistoric sites, a trace memory of early humans from the bronze age displaced by iron age immigrants, hence their dislike of iron. MacRitchie's ideas influenced authors of so-called Celtic Revival literature such as Arthur Machen whose stories *The White People* (1899) or *The Shining Pyramid* (1906) describe human encounters with mysterious races that end in either madness or death. In these traditions the fairy is romanticised as a race apart from humanity that inhabits ancient or liminal worlds within ancient woodlands and burial sites.

Malicious Trees, Death, and Resurrection

Trees are often associated with life, fertility, and renewal. In this story, they are malicious, facilitate death, life-in-death, and are used to imprison, murder, or trick humans. But they are also associated with resurrection. Trees in whatever form are imbued with life as are inanimate and living things in the story because they are all connected under the dominion of John Uskglass. Indeed, as Mr Lascalles declares to Norrell, in Uskglass's time, trees walked about (71). Where the landscape connotes sedimented history, in this story, trees connote deception and death in their incarnations as seeds, books, spells, and monuments. They contain and retain the memory of English magic. Norrell's house, Hurtfew Abbey, was built by John Uskglass and although the house is "solid-looking" (10) the trees in the park are "ghostly-looking" (10). Trees surrounding Lord Pole's house are malicious according to Robert, the footman, they form "an invisible wood growing up around [Lord Pole's] house … [with] … ghostly branches scraping at the walls and tapping upon the windows" (182–3). They tap at the windows and their roots threaten to tear the house down. When Strange meets George III, they are transported to a wood which appeared "sinister, unknowable, *unEnglish*" (463) inferring that Englishness is aligned with certainty.

The certainty of Englishness is challenged in the ways that trees are used to trick and deceive by the Gentleman. Arabella, Jonathan Strange's wife, is sold to the Gentleman for a piece of wood. The Gentleman substitutes the moss oak for Arabella, like fairies substituted changelings for human children, in folklore (Eberley 1988; Purkiss 2000). The moss oak is a piece of oak preserved in a bog and magicked to form an apparition of Arabella. The marsh is a liminal place: neither earth nor water, it is associated with death and eternal life, for marshes preserve the bodies of people who were sacrificed or fell into them. They, therefore, form a direct link to a distant past. The moss oak wanders the moors in the form of Arabella and when brought back to Jonathan Strange, asks him whether he agrees it is his wife. With his agreement, Strange unknowingly gives Arabella to the Gentleman. The use of moss oak to trick Strange refers to its use in ancient Irish folklore "a fairy glance … throws the object into a death-like trance, in which the real body is carried off to some fairy mansion, while a log of wood, or some deformed creature, is left in its place" (Wilde 1887: 52).

Despite their malicious character, England's trees formed a contract with Uskglass and it is contained in a rhyme in which oak trees aid against enemies and ash trees will mourn until Uskglass returns (485). When summoning the Uskglass, Norrell and Strange aim to find him through the sedimented history of the landscape written in the "fruits of the orchard trees", the pears, and apple pips. The orchard trees grew from the pips spat out by Uskglass. The library, too replicates a wood as it is filled with carvings and Gothic arches, "intertwining roots and branches, carvings of berries and ivy" (13). The walls are covered with a design of oak leaves and twigs and the canopy in the ceiling is painted to look like a glade in spring. The books compound that impression looking like dried leaves, leaves being the French meaning of the word 'book'. The library is the place of entry into the real world for Uskglass when the books turn into ravens.

A parallel of the history of English magic written in the landscape is in the writings of the books in Norrell's library and in Uskglass's book. Uskglass's book was consumed by Vinculus's father, Clegg, when drunk. Its contents magically transferred to Vinculus's body making him a living book. Vinculus's hanging body is symbolic in a number of ways: his resurrection later in the story, is reminiscent of the messiah myth and it is reminiscent of the strange fruit of the song inspired by lynching.[1]

The final allusions to trees and death is in their consumption of people and the Gentleman. Lost Hope, the setting in which the Gentleman imprisons and plays with his victims who are entranced, forced to dance night after night, at the Gentleman's pleasure in a grand ballroom. In the television series, tree roots and branches entwine the space and, in the centre, there stands an inverted tree. The inverted tree is similar to a wood or sea henge, surrounded by circles of tree trunks and buried in a freshwater marsh that was eventually submerged and preserved with sea water. Two such monuments were discovered on Holme Beach, Norfolk, 1999, dating to 2049 BCE (see Figure 7). They are only viewable at low tide. It was believed they were used as excarnations, mortuary enclosures in which dead bodies would be displayed so the flesh could be picked from their bones. Thus, the inverted tree was used as an altar and the henge, a gateway to the afterlife. However, when considering it alongside

---

1   Other deaths that reflect the crucifixion are that of Drawlight's body, when shot by Lascalles, which is consumed by trees until "it looked as if he had been crucified" (903). There is also the scene in Faerie in which a wood is filled with corpses hanging from every tree (912) and this is reminiscent of a similar scene in *Excalibur* (Boorman 1981) where Mordred first encounters Sir Percival in a sea of hanged knight's bodies.

fairies, it can connote the world turned upside down and death as an inversion of life, certainly this is the case with the destruction of the Gentleman.

Stephen, the nameless slave, is prophesied to become the King of the strange country by Vinculus. Stephen and Uskglass share similar identities – they are both unnamed slaves. Stephen is an African slave and Uskglass, the Raven King, is also called the Black Knight. Their connection grows even stronger when Strange and Norrell cast a spell to connect all the elements of the landscape "tree speaks to stone. Stone speaks to water ... Uskglass's old alliances still held" (901) which bestows all of the magic in England on Stephen. Stephen uses his power to attack and defeat the Gentleman in the ballroom. In the book he uses millstones, but it is significant that, in the television version, it is the tree that kills the Gentleman, engulfing and consuming him in its boughs. When the Gentleman is defeated by Stephen, Lost Hope disintegrates. In the book, Suzanna Clark relates how, in the collapse of Lost Hope, the mansion becomes a brugh, "This is the world beneath the hill ... The old King is dead. The new King approaches!" (Clarke 2004: 988). Stephen, much like King Arthur, assumes his place as the new King on the ancient throne and resolves to bring discipline to his kingdom.

The image of the Gentleman consumed by the tree, provides several references to myth first in the legend of Merlin entombed in a tree by Numue. Second, in the Green Man, a pagan carving often found in churches and symbolising fertility (see Figure 8). The image proliferated especially in the

Figure 7. Sea Henge at Holme Beach, Norfolk. Image in the public domain.

Figure 8. Green Man or Beast (c.12th century) a pagan symbol often found in churches. St Chad's Church, Stafford, England. Photo taken by the author.

12th to 15th centuries (significantly the time that Uskglass ruled the North). Note the title for this carving is Green Man or Beast – the latter name is given to the Gentleman by Stephen. The image also refers to an older fertility myth of the dying/reviving god described by James Frazer in his epic tome, *The Golden Bough* (1998) in which the ailing king must die to make way for a younger more vigorous successor (much like Uskglass gives way to Stephen Black as the new king).

The darker, more demotic imaginings of the fairy can be identified in contemporary texts in which fairies have their own agendas and operate in liminal and other, "fairy" realms. Many of these representations of fairies are about storytelling and they evoke the Gothic romanticisation of the British landscape. Stories about fairies can tell us something of the concerns of the culture. This story reflects how Britain looks back to, and in some cases, fetishises previous epistemic times of disruption and change. The setting for this story, reflects a time, like our own, when the world was in flux. The British landscape with its malicious fairies reflects this flux. It can reveal how we perceive changes in national identity for fairies act as initiators of magic and chaos in a rational world but by addressing the disruption and controlling the chaos in stories, perhaps we seek to bring order to a world that is disordered.

Nick Freeman

"The New Daughter" (John Connolly, 2004)

The Irish novelist, John Connolly (b.1968), is best known for his "Charlie Parker" thriller series, which began in 1999 with *Every Dead Thing*. The novels blur the boundaries between crime and horror fiction, frequently introducing occult and supernatural elements to striking effect. There has also been, at times, a notable folkloric presence in the sequence, particularly in *The Woman in the Woods* (2018) and its sequel (of sorts), *A Book of Bones* (2019). Elsewhere, his non-Parker fantasy, *The Book of Lost Things* (2006) draws on Robert Browning's "Childe Roland to the Dark Tower Came" (1855) and Angela Carter's *The Bloody Chamber* (1979) in a witty and often dark reworking of fairy tale motifs, settings, and characters. Connolly is a *knowing* writer, one who is always very deliberate in his allusions and his *mis en scène*.

This knowingness is especially apparent in *Nocturnes* (2004), a collection of stories which grew out of a commission from BBC Northern Ireland in 1999. "I had always been fascinated by supernatural stories ever since I was a young reader", Connolly writes, "and I was curious about writing for radio" (Connolly 2004: 487). A number of "nocturnes" were read on BBC Radio 4 in 2000 and 2004, and the book's success gave rise to a sequel, *Night Music*, in 2015. The collection showed Connolly's deep knowledge of his field, from "golden age" ghost story writers such as M. R. James to contemporary figures such as Stephen King and was, he said, "something of a labour of love" (487). Although it contained a Parker novella, "The Reflecting Eye", *Nocturnes* was primarily an opportunity for Connolly to pay homage to his formative influences and to experiment with ideas, situations, and tropes which could be developed or revisited in longer works.

A case in point is the collection's fourth story, "The New Daughter" (2004: 107–22), which imbued a single father's struggles to raise his two children with Connolly's knowledge of Irish folklore. Following the breakup of his marriage, the unnamed narrator buys an old rectory for an absurdly low price after it has stood empty for months and moves to the countryside with his two children. Such bargains are familiar ones in horror fiction and film, and they invariably portend ominous events: Borley Rectory was famously known as "The Most Haunted House in England" during the 1930s. In this tale, the abandoned rectory carries an additional symbolic charge in the implication that Christianity has been driven out by something more powerful, a frequent occurrence in the Parker novels.

The narrator's children are Louisa the more "mercurial" (Connolly 2004: 112) and imaginative one, always "inquisitive and testing of the constraints placed upon her" (Connolly 2004: 110) and Sam, five years younger, who stays closer to his father and is generally more passive. The children's exact ages are not given, but the summary of their parents' marital breakup in the opening pages implies that Louisa is around 11 and Sam is almost 5. The rectory's previous tenant was a woman who worked as an illustrator of children's books and who left behind a number of horrific drawings entirely unsuited to young readers. These depict "pale, half-human creatures with melted features", "narrow oval eyes" and "long tattered wings" reminiscent of "dead dragonflies rotting on a spider's web" (Connolly 2004: 112). She attempted, without success, to burn the images and, we learn later, died in a house fire after escaping the rectory. The narrator wonders if exposure to these nightmarish beings has led her to kill herself, though we are given no objective evidence for this and may simply consider the woman to have been an artist with a darker side to her imagination than was able to surface in her published commissions.

In the 5 acres of land surrounding the rectory there is what is termed a "fairy fort" (Connolly 2004: 112), a mysterious mound 6 feet high and 20 feet across. These "raths or forts" as W. B. Yeats writes, house "land fairies" and "[m]any a mortal has been enticed down into their dim world" (Yeats 1993: 21). Local farmers shun the field containing it and refuse the narrator's generous offer of grazing rights for their sheep, but Louisa is strongly attracted to it, thinking of it as a "fairy castle" (Connolly 2004: 113) and believing that its inhabitants call to her at night. The mound might be seen as a symbolically feminine space, the only one to which Louisa has access in her male-dominated household, and she spends an increasing amount of time there. A crescendo of awful events ensues, for, as W. Y. Evans-Wentz recorded in 1911, many in rural Ireland believed that "[i]f a house happens to be built on a fairy preserve, or in a fairy track, the occupants will have no luck. Everything will go wrong. Their animals will die, their children fall sick, and no end of trouble will come on them" (1911: 38).

First, Molly, Louisa's childhood doll, disappears and is replaced in her bed by a horrible effigy with dandelion leaves for wings. A large but sickly spider is imprisoned in it, withering slowly away.[1] The narrator then has a vision at the mound of a creature from beneath it, something that "has spent too much time away from the light" with a vile mouth, "set in a permanent grin" and "ruined, wasted wings" (Connolly 2004: 117). He discovers Molly's head buried nearby, the doll's hair a squirming mass of beetles and worms. He also finds scratches on the outside of Louisa's bedroom window and wood torn from its frame. Louisa begins to sleepwalk.

Connolly's familiarity with the conventions of horror film and fiction are obvious from the story's next episode, in which the narrator has to leave the children in the overnight care of his housekeeper while he goes to London on "unavoidable business" (Connolly 2004: 118). That the housekeeper is called Mrs Amworth and therefore shares her name with the eponymous female vampire of E. F. Benson's 1922 short story, is another wink to the horror aficionado. We do not see events at first hand, Connolly instead using Sam to provide a halting and partial account of how his sister asks him to play at the fort at night. "You like playing with her", the narrator says, "conscious suddenly that while this might once have been true, it was true no longer" (Connolly 2004: 119). Sam's teddy bear disappears; the narrator finding its head buried near the mound. Louisa becomes uncommunicative, silent, wary, and her appetite changes so that she eats only meat and disdains vegetables. When her father finds her fiddling with the lock on Sam's bedroom window, a row ensues, with him shaking her by the shoulder. "I looked into her eyes and saw something red flicker in their depths", he says, her eyes seeming narrower and

---

1 Such descriptions recall the fairies fashioned by the artist, Tessa Farmer (b.1978), the great-granddaughter of Arthur Machen: www.TessaFarmer.com.

more slanted than before and more like those of the creatures in his vision. "Don't touch me", Louisa whispers, her voice tainted by a "hoarse, filthy aspect". "Don't ever touch me again" (Connolly 2004: 120). With familial tensions intensifying by the moment, the story comes to its unsettling finale in which Louisa appears beside her father's bed, announces, "I am your new daughter" and warns that he will not always be able to keep Sam safe. "Some night you'll be careless, and then you'll have a new son, and I will have a new brother" (Connolly 2004: 122). The narrator puts the rectory up for sale and waits anxiously to flee with his now traumatised children.

This ending can be read in several ways. If the fairy fort really does house some lost race of malign winged supernatural creatures who are able to substitute Louisa for a child of their own, the story is one of outright horror. If, however, the fairy element operates on a metaphorical or symbolic level, it serves to represent the father's incomprehension at the "otherness" of female adolescence. "New" can, after all, refer to a person's reinvention of themselves (a "new" me). As the narrator says, when a father tries to bring up a daughter alone, "there will always be some part of her hidden from him, unknowable to him" (Connolly 2004: 109). And how rational and reliable is the narrator of these extraordinary events? He is, after all, worn thin by an obviously demanding job and a messy separation from his former wife, and is trying to bring up two children in different age groups in a five-bedroomed rectory seemingly far from any other habitation. By moving to the country, he has also cut them off from their school friends and wider family. His bad dreams and waking anxieties are hardly surprising given the strain he is under, and rational explanations could be found for, say, the hitherto unnoticed damage to the windows or Louisa's destruction of her brother's teddy bear. Connolly's teasing ambiguities are irresolvable in a first-person narrative. As he says of short fiction: "The pleasure of the supernatural short story lies in not having to explain anything. They're just glimpses, a momentary lifting of the veil" (Mynhardt and Johnson 2017: 161).

Unfortunately, John Travis's screenplay for the film version, *The New Daughter* (2009), starring Kevin Costner, played the supernatural card rather too overtly and robbed the story of these appealing ambiguities, offering a justifiably concerned (and gun-toting) father instead of the more conflicted character of the original. Transplanting it to the United States also detached it from its original cultural context: whatever lives in the mound in Connolly's original is firmly of the old world, not the new.

"The New Daughter" is a story which makes artful use of fairy lore to explore the end of childhood and the troubling aspects of father–daughter relationships, while always being aware of the conventions of the modern horror story. It adapts the "cursed child" motif which runs through horror fiction from Henry James's *The Turn of the Screw* (1898) to novels and films of the 1970s such as William Peter Blatty's *The Exorcist* (1973) and Frank de Felitta's *Audrey Rose* (1975) which introduced some form of demonic possession and on to more contemporary anxieties about children and childhood (see Hendrix 2017). It would, of course, be easy enough to address these through the prism of realism, but Connolly's references to fairies allow an additional perspective, as well as perhaps winning a wry smile from parents of teenagers. The horror of "The New Daughter" resides in part in the idea that the narrator has "lost" Louisa, not because she has disappeared or been abducted (frequently the starting point for one of Parker's investigations) but because she has been changed – she may have Louisa's external appearance, but her personality is suddenly radically different. Whether this change arises from her encounter with the things in the mound, or whether it is a natural consequence of Louisa's adolescence and the tormented evolution of her new selfhood, we are left with a terrified but uncomprehending father who is all-too aware that Louisa is right. One day, meek, childish Sam will be a quite different creature, too.

The most important fairy aspect of the story is its reworking of changeling legends, found throughout the British Isles since the Middle Ages and which, as Carolyne Larrington says, are often "distressing reading" (2015: 216). In these tales, fairies, covetous of human children, steal them and take them off to fairyland (often portrayed as the world beneath a hill or mound), replacing them with surrogates. Sometimes these replacements are blocks of wood or bundles of sticks which, through magic, are made to seem briefly like the abducted baby. When the glamour wears off, the "baby" dies, and the substitution goes undiscovered. On other occasions, the human child is replaced by a fairy which does not thrive and eventually expires, or, perhaps worst of all, by an outcast fairy which, as Katharine Briggs writes, is "of no more use to the fairy tribe and is willing to lead an easy life by being cherished, fed, and carried about by its anxious foster-mother, mewling and crying for food in an apparent state of paralysis" (Briggs 1976: 70). Older children were abducted to improve the fairy stock, or simply because the fairies desired their beauty, perhaps the fate of Louisa. We should be thankful that the story ends before her father is able to take any action to reveal whether his daughter has been substituted by something from within the mound, as such tests and trials could lead to cruelty, injury, and death. As Angela Bourke has detailed in *The Burning of Bridget Cleary* (1999), as late as 1895, in what became known as "The Clonmel Witch Burning", a young Irishwoman was killed by her husband and family in the mistaken belief that they were driving out a changeling by fire and hot iron.

The changeling myth plays out through the abduction, mutilation, and replacement of Molly, the doll which was Louisa's nocturnal companion as a baby. Louisa denies having anything to do with Molly's burial (and is upset by the idea that she was "lost beneath the earth", as the narrator puts it [Connolly 2004: 117]), but the rejection of the emblematically innocent Molly in favour of the more grotesque stick effigy works to imply that Louisa too has been replaced while also suggesting the familiar scenario in which a teenage girl radically changes her image from childhood "prettiness" to something more challenging, even, in the opinion of her parents at least, deliberately ugly. The changeling, like many teenagers, has no desire to sacrifice itself to please or coddle those who care for it, but it is difficult for the narrator to understand or accept the changes in Louisa's behaviour, and he prefers to attribute them to supernatural intervention, even though this is surely a far more disturbing and unlikely explanation than a shift in his daughter's hormones. Her silences, her moodiness, her fussy eating, her growing distance from her "baby" brother, and the obvious tension between herself and her father sound like an unexceptional familial experience as much as a source of horror and fear.

Despite its brevity, "The New Daughter" is a subtle and unsettling story which sets up a fascinating opposition between a "modern", if somewhat dysfunctional family, and the forces of legend and superstition. The opposition of modernity and its disregarded Other is, of course, a staple of horror and gothic; the repressed never ceases to return and nothing remains forgotten or buried for long. Connolly's story however offers a fresh version of this perpetual contest, not least because, as fairies have become ever more twee and Disneyfied, "The New Daughter" returns us to the more dark and troubling aspects of the fairy world.

Lorna Piatti-Farnell

Carnival Row (Travis Beacham and René Echevarria, 2019–Present)

Carnival Row is an American television series, created by René Echevarria and Travis Beacham, and released on Amazon Prime. Often considered to be a mixture between fantasy and noir genres, it leans distinctly on a blend of steampunk and Gothic aesthetics, especially in terms of fashion, style, and weaponry. The series focuses on an imaginary world where mythological Fae creatures have fled their native land of Tirnaroc as an outcome of colonisation and war by encroaching human factions. Dispossessed and often desperate, the Fae have primarily moved into human areas – such as the city of The Burgue – as either immigrants or refugees. The Fae community is composed of several races, with Faeries and Fauns being the most prominent in human contexts. The plot of *Carnival Row* focuses particularly on the narrative exploits of Rycroft "Philo" Philostrate (Orlando Bloom), an half-Fae Inspector of The Burgue Constabulary, and his lover, Vignette Stonemoss (Cara Delavigne), a Faerie with wings – who is also commonly referred to by the humans with the derogatory term of "pix". Taking this context as a point of departure and focusing primarily on Season 1 of the show, this chapter explores intersecting notions of ecologies, urban identities, Otherness, and racial politics, with an aim to analyse *Carnival Row*'s response to evolving cultural discourses in the 21st century. Indeed, operating within a suggestively Neo-Victorian context of history and politics, the series is focused on a variety of concerns that tacitly recall popularised interpretations of the Imperial era – including gender oppression, immigration, environmental issues, and (post-)colonialism – that "still resonate" with contemporary audiences (Pedro 2021: 245).

As a result of the mass exodus from Tirnaroc in previous decades, The Burgue –the eponymous capital of the Republic of the Burgue – is now not only one of the most prominent centres of human culture in the Mesogean continent, but also the focal point and primary destination for the Fae, who continuously seek refuge and settle in the city. In the wider world, The Burgue is continuously at war with The Pact, another human superpower who also vies for the possession of territories, including those that have been conquered in the lands of the Fae. Within The Burgue, the particular area of Gloamingside, with its specific neighbourhood of Carnival Row, is now a central location where the Fae live and work. As the Fae are not permitted to become involved in human activities to make a living, they often resort to either manual labour or businesses of a sordid nature – including brothels – to make ends meet. One can see here how the show tacitly constructs the socio-political exchange between the humans and the Fae as recalling "coloniser" and "colonised" statuses. As Samira Sasani suggests, "the relationship between the colonizer and the colonized is based on an exchange where the identities of the colonizer and the colonized (the Other) are mutually constructed" (2015: 435). In the case of *Carnival Row*, the humans develop their coloniser identity as based on frameworks of superiority, constantly abusing the Fae and negating them freedom and opportunities. For their part, the Fae exist within the "colonised" identity by being not only submitted, but also by being socially and culturally shaped by colonial power. This is highly reminiscent of Edward Said's well-known and influential analysis of the colonial Other, where the coloniser perceives the colonised as inherently inferior (Said 1978).

The city of The Burgue is the political centre of the Republic, and where Parliament is held. Especially in Season 1 of the show, two major parties are shown competing for the control of the city: the conservative Hardtackers party, led by Ritter Longbane (Ronan Vibert), and the more progressive Commonwealth Party, led by Absalom Breakspear (Jared Harris). Discussion over the place of the Fae immigrants and refugees in The Burgue – including the rights of the Fae – are often a contentious point of debate for the two parties. The races of the Fae kind are collectively known by the humans as "otherkin"; however, the derogatory term "critch" is most used in the series by the inhabitants of The Burgue, signaling the deep-rooted hatred and mistrust that the humans hold for the Fae who have settled in the city. Far from being the safe haven that the Fae had hoped for, The Burgue is a city filled with "racism, alienation and decivilization" (McGregor 2021: 267). As part of the contemporary contextualisation of racial and ethnic politics in the show, the status of the Fae as mainly refugees strongly recalls the right-wing politics that have recently mongered an unfounded "fear" of immigration as a destabilising presence in the real world, especially in Western countries such as the United States and the United Kingdom. The Fae are often identified by the humans (paradoxically) as a violent and untrustworthy group. Here, the immigrant – or refugee – is constructed as the "absolute negative" Other (Khair 2009: 4), through the evocation of dread and perceived acts of terror.

The imagery and dark colour palette used throughout *Carnival Row* to depict The Burgue mark it as a smoky and filthy place, which is unsurprisingly also proven to be a highly corrupt and corruptible entity. The visual similarities between popular literary and media depictions of Victorian London here are difficult to miss, and this is instrumental in placing the series within a cultural framework that echoes the metaphorical use of Imperial politics in our contemporary era (Pedro 2021). Indeed, the visual representation of The Burgue communicates its role in the series as a "cultural metaphor" (Haapala 1998); indeed, this is a function that the city often occupies in historical fiction. Within the cultural confrontations between the humans and the Fae, The Burgue is particularly of note for standing in cultural and spiritual opposition to the land of Tirnaroc. Where Tirnaroc – before the human conquest – was imagined as a lush land filled with natural beauty and defined by positive interactions between creatures, The Burgue is characterised by imposing buildings, technological structures, and dangerous districts, signifying both the impenetrability of the humans as a society and the impact that "progress" has had on their evolving politics and moralities. The Burgue is a city of ambiguities, and, even in its most grand neighbourhoods, aptly recalls what Walter Benjamin and Asja Lacis (1978) termed "the porous city", filled as it is with balconies, arcades, courtyards, staircases, and windows that mesh and join together in the inseparability of the urban collectivity. Far from evoking notions of connectedness and potential openness (Wolfrum 2018), however, The Burgue is an impenetrable, fragmented, and paradoxically impermanent entity, and promotes ecological barriers that rely on the instrumentalisation of difference to create disorder. In this way, The Burgue becomes a liminal space that promotes almost dystopian aspects of the cultural and racial politics that define the human vs Fae dichotomy (see Figure 9).

In contrast to the disaffecting urban structures of The Burgue, the land of Tirnaroc is the centre of Fairy ecologies. Tirnanoc is a continent located on the East side of the Great Main Ocean. The region comprises of the ancestral homelands of the Fae and is the place of origin for all their customs and beliefs. It is not difficult here to note the conceptual similarities between Tirnaroc and Tír na nÓg, the supernatural realm of Irish mythology, which is considered to be the place where the Fae come from (Koch 2006). *Carnival Row* suggests that Tirnanoc is comprised of a variety of nations and kingdoms, each associated with a particular race of the Fae. While the history of Tirnaroc is rich and long-spanning into millennia past, recent centuries have seen its reach dwindle, as the encounters with humans, and the impacts of human colonisation, increased. Indeed, it is mentioned on multiple occasions that various human powers have been crossing the ocean for some time, claiming territories and dispersing the Fae as a result.

While Tirnaroc has become an unstable place for the Fae to inhabit after the conquest, the memory of its glory is well-rooted and is perhaps most

Figure 9. Rycroft "Philo" Philostrate (Orlando Bloom), the half-human/half-Fae Inspector, patrols the streets of Carnival Row in The Burgue. *Carnival Row*, created by René Echevarria and Travis Beacham (Amazon Prime Video, 2019–23).

symbolically embodied in the much-protected surviving Tirnanoc Library, as shown in the episode "Kingdoms of the Moon". The Library, comprising of some of the oldest tomes from Fae history and knowledge – some being thousands of years old – is located in a hidden cave in the Tirnanese highlands, in the Kingdom of Anoun, and represents one of the few glimpses that viewers get of "the home world of the faeries" (Moltenbrey 2020: 42). The Library is safeguarded by Vignette, who has sworn to protect the valuable volumes in the cave. It is made clear that the Library is an essential part of faerie history, and representative of their essence as a multifaceted people. Particularly of note here is the fact that the Library is concealed deep in the rock, and made to be – one might argue – part of the mountain itself. Mysterious and hidden, the Library is projected as the beating heart of the Fae in Tirnaroc. Its non-intrusive presence metaphorically signals the reciprocal relationship with nature that the winged Fairies in particular appear to have. Often projecting themselves as guardians of the natural world, the Fae also openly owe their livelihood to nature. The fact that the core of their knowledge is entrusted to the mountain speaks loudly about the importance of ecological exchanges in the faery world.

It could be argued here that the series seems to be gesturing towards a figurative rendition of Indigenous knowledge in order to construct the environmental politics of the Fae world, and especially so in the case of the winged Faeries. When considering approaches to nature, it would be unwise to generalise and assume that all Indigenous worldviews are "the same". Some recurring similarities, however, can be traced. Indeed, while obviously differing in several ways – primarily connected to specific and intricate notions such as "place" – Indigenous views often hold an understanding of nature as having its own intrinsic value and life force (Niigaaniin and MacNeill 2022). In various Indigenous views, humans are part of nature, while also being entrusted with its care through ideas of partnership, custodianship, and guardianship. In Indigenous systems, land is commonly not "held by individuals"; instead, a relationship is formed with the land by "a variety of relational subjects", including "non-humans" (Trosper 2022: 87). This cooperative understanding of the relationship between humans and the environment is founded in a mutual exchange. In *Carnival Row*, the Faeries' approach to nature seems inspired by Indigenous worldviews, at least as far as the relationship with nature is concerned. Of course, the show puts the idea of "Fae" in contrast with that of "human"; nonetheless, given the particular ideological framework that can be identified as part of ecological politics, it would not be too ambitious to see the metaphorical understanding of "the Fae" as a race of Indigenous people, and the humans of both The Burgue and the Pact as the forces of Western colonisation. This notion is reinforced by the fact that people in The Burgue refer to the Fae as "otherkin", echoing the disturbing colonial discourse that identified Indigenous populations as "Other". In the context of the Fae in *Carnival Row*, however, it is made clear that it is in fact humans who are "the Other": a destructive force that not only to lay claim to the culture and memory of the Fae as a collective race, but also shows disregard for the natural environment by causing ecological destruction in Tirnaroc lands. This point is indeed reinforced when, later in the series and after the Tirnanese highlands have been conquered, Vignette visits a museum in The Burgue that houses the remaining contents of the Tirnaroc Library. Her shock at seeing the precious faerie artefacts displayed in such fashion causes her to become visibly upset and speaks closely to the cultural appropriation and identity erasure that is intrinsic to colonial policies.

Ethnic and racial politics are indeed present throughout *Carnival Row* and employ the perceived and reinforced differences between humans and Fae as metaphorical conduits for uncovering real-world sociological preoccupations. While the seeming contrast between "Indigenous" and "coloniser" appears clear in the show, what can also be identified is a distinct representation of the politics of racial difference in terms of "Black" vs "white" dichotomies, under the constructed terminologies of "Fae" and "human" respectively. This is perhaps most clearly exemplified in the figure of Philo. The son of a human father and a Fae mother, Philo is regarded as a "half-blood", and therefore occupies a despised category within the racial and ethnic structure of The Burgue. However, Philo physically presents as mainly human, as his undeveloped wings were amputated while he was still a baby, in an effort to make him fit in the human world. Without the presence of his faerie wings as a marker of difference, Philo is able to "pass" as a human in The Burgue, and able to pursue the detective career that the Fae would be precluded from having. Philo's situation carries deep-rooted historical echoes. Indeed, the notion of "passing" has a long history in both fictional and sociological contexts and has been a central part of the American landscape since the 19th century (Rummell 2007: 1). The term passing refers to "a practice conducted by individuals belonging to minority or oppressed groups that pretend to be members of the privileged culture to enjoy their same rights and escape marginalization" (Pedro 2021: 248). In the series, Philo goes through great efforts to not reveal his half-blood status; once he becomes aware of his background, he also becomes fearful of discovery, and even goes as far as refusing medical help when injured, in order to keep his Fairy heritage concealed. Both his behaviour, and the decision that Philo's parents made to cut off his wings in infancy, place him within a "passing framework", as the narrative uncovers the prejudices and humiliations that the Fae have to suffer in the human world.

As humans and Fae are considered to be two different races, it is not difficult to see the metaphorical rendition of real-life racial politics in Philo's plight. As Kathleen Wehnert argues, the term "passing" is most employed to refer to people "crossing the colour line from Black to White to surpass racial obstacles" (2010: 1). As the history of this practice is interwoven with American politics, it is not difficult to see those nuances clearly overlaid on Philo's predicament in the series, even if the "races" in question are overtly those of human and Fae. Read as a "passing narrative", *Carnival Row* clearly delves into a familiar account where mixed-race individuals distance themselves from their Black identity to "fully embrace" the white community and its privileges (Pedro 2021: 248). Later in the story, and once it is publicly discovered that Philo is in fact a "deplorable" half-blood, he is quickly ostracised, and suffers the treatment that humans reserve for the Fae. However, because of his mixed-race background, Philo is also swiftly branded as "the Other" by both communities. As he attempts to join the Fae in Carnival Row instead, he is also treated with mistrust, signalling the issues of belonging and acceptance that are impossible to extricate from discourses of racial difference.

As the cultural conflict between humans and Fae unfolds in The Burgue, it is made clear that *Carnival Row* employs physical differences as a base for racial and ethnic discrimination. Indeed, the Faes' bodies are fetishised as representative of their behaviours and difference, and become synonymous with their inferiority within human contexts. The series appears seemingly keen on placing an emphasis on the bodies of the Fae as central parts of their characterisations, leaning on an "exoticizing representation of these mythological creatures" (Pedro 2021: 255). It quickly becomes apparent in the narrative that the Fae are expected to somehow hide their physical differences, as a form of corporeal and socio-cultural repression for their abilities. Indeed, many central segments in the plot of *Carnival Row* are focused on an exploration of the Fairies' wings. When female Fairies enter the service of wealthy families of The Burgue as maids – a common profession for them – they are forced to conceal their wings in specially designed busts; this not only forcibly suggests their nature to be "humans" to onlookers, but also "removes" the elements that could suggestively and dangerously expose them as superior – that is, the ability to fly. Female Fairies' wings are also fetishised in a sexualised manner, as – when not concealed – they are treated as a point of dangerous attraction. This is made evident in the depiction of the Fairies prostitutes working in the brothels of Carnival Row, whose bodies and wings are placed fully on display for potential customers. This is particularly visible in Episode One, "Some Dark God Wakes", as part of the contextualisation of the differences between Fairies and humans. Here, the wings are treated as an embodiment of not only difference but also promiscuousness, an aspect that immediately places the Fairies on the bottom rung of the social scale. As Dina Pedro argues, Fairy prostitutes "are not bound" by "social constraints", and their Othered bodies are therefore "exoticized and hypersexualized" (2021: 255). The sexualisation of Fairies wings is further explored in "Kingdoms of the Moon" (see Figure 10), when, during sexual encounters between Vignette and Philo, her wings are shown to glow in shades of green and blue, as she reaches the pleasure climax.

The fact that the majority of Fairy prostitutes in *Carnival Row* are female is also indicative of the series' engagement with gendered politics. Here, a clear contemporary and historical sense of "racialized Otherness" is "displaced onto the Fae" (Espinoza Garrido 2020: 221), and particularly those who are female. In spite of the fact that the majority of characters in the series are played by white actors, it is possible to argue that the narrative tacitly points to the representation of female Fairies as emblematic for the stereotype that identifies people of colour, and especially women, as highly promiscuous. Indeed, the visual composition of the fairies' bodies seems to expose the humans' reliance on what Felipe Espinoza Garrido recognises as the "biologization of racial difference", which encourages the enactment of racial disparity and supremacy (2020: 220). While this discourse finds its roots in 19th-century politics – and has been particularly influential in British and American societies – it is not

Figure 10. A closeup of Vignette's (Cara Delavigne) Fairies wings in the episode "Kingdoms of the Moon". *Carnival Row*, created by René Echevarria and Travis Beacham (Amazon Prime Video, 2019–23).

difficult to see how its utilisation in *Carnival Row* also proposes a deep-rooted critique of racial politics in the 21st century.

In similar fashion, the Fauns' bodies – which display hooves and horns – are shown as a further corruption of accepted human physicalities, as it is also suggested that this race of the Fae hold preternatural abilities, like an enhanced sense of smell. This representation is particularly relevant in relation to the Faun character of Agreus Astraryon (David Gyasi). His animal-like appearance is treated with disgust and disdain by the humans. His physical qualities quickly become synonymous with racial difference, which immediately brands the Faun as a "barbarian", in spite of the fact that Agreus is not only incredibly wealthy, but also well-mannered and highly educated. It is worth mentioning here that Agreus is one of the few characters played by a Black actor in *Carnival Row*, adding a further tangible layer of racial politics to the representation. Later in the series, Agreus is suggestively shown as having an affair with his human neighbour, Imogen Spurnrose (Tamsin Merchant), in what is openly perceived as a despicable, shameful, and even "unnatural" act in the eyes of the humans. It is not difficult here to see a critique of mixed-race couples in countries like the United States, where (in the pre-Civil Rights era) inter-relation relationships were not only frowned upon, but actually banned in many jurisdictions (Romano 2009). This was particularly true of relationships between African American men and white women.

In conclusion, one can see here how the physical and cultural representation of the Fae/human divide in *Carnival Row* emerges as both "raced" and "gendered" (King 2014: 173), as the show reinforces and attempts to decolonise the politics that have historically defined notions of racial Otherness, and the perceived opposition between "natural" and "unnatural" in cultural discourses.

Part III

Daughters, Mothers, and Godmothers

The Muses

Gemma Files

Jenny Wise and Lesley McLean

Peter Pan (Clyde Geronimi, 1953)

Introduction

Tinker Bell was first introduced to audiences through J. M. Barrie's *Peter Pan* play (1904) and *Peter and Wendy* novel (1911). Throughout the theatrical productions, Tinker Bell did not assume a physical form, instead she appeared as a flash of small light zig-zagging around the actors on stage. This visualisation embodied Barrie's text, where Mrs Darling describes Tinker Bell as a "ball of light ... like a flame had escaped from the fire, not as big as your hand, but [she] darted about the room like a living thing" (cited in Meyers et al. 2014: 102). While the plays delighted audiences (including Walt Disney in 1913), Tinker Bell did not achieve fame in her own right until she was "humanised" by Walt Disney in the 1950s when the Disney film *Peter Pan* was released on 3 February 1953 and "was an immediate commercial and critical success" (Barros 2007: 1). Originally a Disney Princess (although no more), Tinker Bell has been referred to as one of Disney's "mascots" (McClintock 2014: para. 9). This chapter explores the trajectory of Tinker Bell from a character in a play to being the unofficial mascot of one of the largest corporations in the world.

From Theatre to Disney Movie Star

While J. M. Barrie first created the stage play of *Peter Pan* in 1904, he encouraged different versions of his story to be created. For example, he allowed numerous authors to write different novels of the story aimed at different age groups, each time illustrated differently. According to Bell (2016: 82), there have been "45 different publications of *Peter Pan*; 22 of these *not* authored by Barrie. Of these 22, five are published by Disney". While there are 45 different versions of the story of *Peter Pan*, there is one definitive image of 'Tinker Bell' that comes to mind when her name is heard: the image created by Disney.

In the 1953 Disney adaptation of *Peter Pan*, Tinker Bell assumed a more prominent role in the narrative. Importantly, rather than remaining a flash of light that indicated "lawless, hedonistic, childlike behaviour" (Meyers et al. 2014: 102), Disney's Tinker Bell becomes "humanised", sculpted on the ideals of beauty and femininity of the time. For example, it is believed that Tinker Bell was modelled off the likes of Marilyn Monroe or Margaret Kerry (Chambers 1966: 51). As such, through "modernizing Tinker Bell's 'brand' of femininity, Disney was appealing to specific social, political, and cultural phenomena of the time" (Meyers et al. 2014: 104). In particular, "Tinker Bell's single, working-girl status, her exotic otherness, and her implied sexual availability and desire for Peter Pan all stood in stark contrast to the dominant family values of this era" (Meyers et al. 2014: 105–6).

Unsurprisingly, Tinker Bell's debut appearance in a physical form was highly controversial. Because she was modelled after the "beauties" of the time, her appearance has been described as "a mute, buxom, platinum-blonde, butterfly-sized creature (reportedly modelled after Marilyn Monroe) that communicates with a tinkling sprinkle of magical pixie dust" (Davis 2008: 20), and more controversially as "an elfin sex symbol" (Chambers 1966: 51). Her depiction as "mature" and "scantily clad" was considered inappropriate by some for a children's cartoon (Arcus 1989: 294). Further, her first solo appearance in the film, landing on a hand mirror and admiring her bottom, caused consternation by some parts of society because of her self-awareness and appreciation. This image has become iconic and continues to be used to market products (see Figure 11).

Peter Pan (Clyde Geronimi, 1953)

Figure 11. The image reflects the original scene of Tinker Bell landing on a hand mirror and admiring her bottom in *Peter Pan* 1953. Photograph taken of *Tinker Bell* diffuser made by Short Story, in collaboration with Disney. Reproduced under Fair Use.

In reality, the likeness of Tinker Bell as a "buxom" creature interested in her own looks had already been described by Barrie himself: "It was a girl called Tinker Bell exquisitely gowned in a skeleton leaf, cut low and square, through which her figure could be seen to the best advantage. She was slightly inclined to *embonpoint*" (cited in Bell 2016: 85). And while there was anxiety about her juxtaposition against traditional values of the time (the importance of the family unit; the role that women should assume, etc.), the magic of Disney was reaffirmed with the children's return to the safety of home and their "normal" roles – thus, Peter Pan and, importantly, Tinker Bell could provide a magical escape from reality, which while enjoyable, reinforced the desire for dominant values of society at that time (Meyers et al. 2014: 105–6).

Disney's New Mascot 1954–2000

Since the 1920s, the franchising of the Disney empire has always been a consumer paradise – every aspect of Disney is merchandised and marketed to a wide consumer base, from babies to adults, and all genders. In addition, the Disney entertainment agglomerate includes theme parks, publishing, movie studios, retailing, cruise ships, the Buena Vista Internet group, and a major television network (Dholakia and Schroeder 2001), providing unlimited opportunities to market Tinker Bell as a Disney mascot.

Disney's *Peter Pan* was released at the height of the baby boom culture of consumption (Meyers et al. 2014), providing Disney with the opportunity to provide licensed Peter Pan toys, dress-up clothes, storybooks, records, and board games. Capitalising on this growing consumer market and money for leisure activities, Disney introduced an animated Tinker Bell onto their television channel (first aired in 1954) to "open" each episode. Tinker Bell appeared in the opening credits to shower magical sparks over the screen with her wand, and in April 1958, she was given her own episode to take audiences on a guided tour of Magical Kingdom. Disneyland, California, also opened a 'dark' Peter Pan ride that provided audiences with an immersive, multi-sensory experience in 1955, with versions of this ride now featuring in Disney theme parks in Tokyo, Florida, Paris, and Shanghai.

The use of Tinker Bell to open Disney's television series evolved into the use of a "real-life" Tinker Bell at the Disney theme parks. In 1961, 71-year-old circus aerialist, Tiny Kline, was hired to perform as Tinker Bell while soaring above Fantasyland on summer nights. In the 1965 Disneyland Souvenir Book, the night-time performance was described thus:

> Every Summer evening at 9, fireworks cascade a shower of color over Disneyland. At that hour, Tinker Bell "flies" across the Magic Kingdom – down from the Matterhorn and

high over Fantasyland – drawing the curtain on daytime fun ... and shining the footlights on nighttime magic (cited in Strodder 2008: 169).

Tinker Bell continued to feature in the nighttime displays until 2022, however, improvements in technology meant that rather than gliding straight down, Tinker Bell could move back and forth, and while the fireworks are occurring. In 2023, Tinker Bell was replaced by Baymax as part of the 100th anniversary of the Walt Disney Company evening fireworks over Disneyland (Libbey 2022), there is uncertainty whether this is a permanent replacement, or just for the celebrations.

In May 1987, Disneyland (California) introduced 'Disney Dollars', which was quickly rolled out to Disney World and other Disney Stores in America. While the designs of the notes have changed to keep up with trends, according to Strodder (2008), the front of each bill has always featured a small drawing of Tinker Bell, demonstrating the enduring fan appreciation of the fairy. Tinker Bell has her own 'store' in Fantasyland, Disneyland; a Storybook Land Canal boat named after her; as well as actresses representing her in parades and character 'meet and greets' around the world.

Rebooting the Magic, Tinker Bell, and Disney Post-2000

The 'magic' of Tinker Bell has endured, and Disney consistently re-imagines her to attract new and ever-expanding audiences. For example, in 2005, Tinker Bell became a lead character for the Disney *Fairies* franchise, and in 2008, she was given an "origin" feature film *Tinker Bell*, which quickly led to six more films (the last released in 2014; most going straight to DVD; however, the popularity of the franchise saw the last few films making it to the cinemas first). The 2000s reboot of Tinker Bell and the ensuing franchise (clothes, dolls, Manchester, books, costumes, etc.), earned Disney over US$335 million, which reportedly inspired further high-grossing movies such as *Tangled* and *Frozen*, kept the same likeness of the 1953 Disney *Peter Pan* film – she was palm-sized, sported a high blond bun, green dress and was "slightly voluptuous" (McClintock

2014: para. 9). Despite being palm-sized in the movies, Tinker Bell has been reproduced as a 'standard' doll, allowing children to play with her as they might the more traditional Disney Princesses (see Figure 12). While keeping Tinker Bell "traditional", the other fairies were "ethnically diverse", possessing "empowering skills" (McClintock 2014: para. 9), that enabled the movies to appeal to a wide range of audiences.

Figure 12. Photograph taken of Disney *Tinker Bell* 'classic doll' on sale in Australia in 2023. Image used under Fair Use.

The reboot provided Disney with an opportunity to contextualise, not just Tinker Bell, but the history of fairies in general. According to Meyers et al. (2014: 108):

> Earlier print and screen incarnations of magical figures of myth and fairy tales did not require a back story; they simply existed. In the second half of the twentieth century, however, readers and critics began to evaluate the quality of children's fiction in terms of characterization rather than plot. The shift toward valuing complex characterization is evident in the development of Tinker Bell's character, upon which Disney has now conferred a life story that realizes her full personhood. ... Tinker Bell's story is no longer about a single, sexualized fairy but rather about the state of being a fairy.

The success of Disney *Fairies* has seen Tinker Bell named "Honorary Ambassador of Green" by the United Nations Department of Public Information in 2009 (States News Service 2009). Thus, the franchise was being used to promote environmental issues to young audiences, and conforms within Disney's desire to reflect and shape shifting societal values.

As a part of this, the buxom blonde Tinker Bell's personality is expanded past the original 1953 Disney notion of the fairy being preoccupied with her looks (and jealous of other girls, particularly those who secure Peter's attentions), to reflect modern-day "tween" behaviour. Tinker Bell is jealous, head-strong, and rash; but also loving, compassionate, a fierce friend, and above all, wants to help and "fix" things. As such, the Disney Tinker Bell is depicted as being strong but emotional (with a "dark" streak) enabling merchandise slogans of "Spoiled to Perfection," "Mood Subject to Change Without Notice", "Tinker Bell: Prettier Than a Princess" and "Dark Tink ... the bad girls side of Miss Bell that Walt never saw" (Orenstein 2006: para. 50). As such, Tinker Bell offered a particular generation a role model, that, while being 'good', could also have a mischievous side with attitude – transforming her into a relatable icon for the young and old.

The popularity of Tinker Bell at Disney may be waning. As identified above, there have always been concerns about her emotional reaction throughout the story to other females, and in particular her jealousy, and her apparent obsession with her looks (standing on a mirror to admire her bottom). Disney is a family-orientated empire that relies on reinforcing emerging societal values, and current speculation indicates that Tinker Bell may not be a mascot for Disney for much longer, as she seems to be juxtaposing emerging social values.

For example, prior to the covid pandemic, Tinker Bell had her own grotto in the Town Square Theatre in Magical Kingdom (Disney World, Orlando) where visitors could meet Tinker Bell and have a photo with her. While other meet-and-greet characters slowly returned in 2021–2022, Tinker Bell's "experience" was removed from the website and app. Speculation ensued that her

Figure 13. In this series, Tinker Bell is given prominence by appearing as mostly silver, along with other iconic characters such as Mickey, Minnie, Donald Duck, Dumbo, Simba, Winnie the Pooh, Alice in Wonderland, Elsa, and Moana). Photograph taken of Disney *Tinker Bell*, no. 25 of the Woolworths Disney 100 Wonders card collection. Reproduced under Fair Use.

removal was based on a review into all Disney characters and experiences by Disney's "Stories Matter" team. The team labelled Tinker Bell as "problematic" and to be "marked with caution" because she is "body-conscious" and "jealous of Peter Pan's attention", initiating theories that Tinker Bell meet and greets would not return to Disney World (Bosacki 2022: para. 9). Indeed, the Disney World website advertises that visitors can meet Peter Pan (and shows Wendy in the picture, but not Tinker Bell). Despite these speculations, Tinker Bell meet and greets still occur in the Disneyland Resort in California; and she appeared briefly at Epcot in January 2023 for meet and greets and then had her own area (*Tinker Bell's Fairy House Garden*) in the 2023 Epcot Flower and Garden Festival (Disney World) demonstrating an ongoing relationship with Tinker Bell as an unofficial mascot.

Further, in 2021, a "life-size" (tiny) Tinker Bell meet and greet was play tested at Disney California Adventure, and rumours continue in 2023 about the possibility of this being the new "meet and greet" experience for guests. This technology allows a life-size tiny Tinker Bell to appear in a lantern and then fly around the room. Whilst in the lantern, her voice was amplified, enabling her to talk to guests, including addressing children by their names (Michaelsen 2023). Tinker Bell also featured heavily in the Disney 100 celebration merchandise, appearing on a range of merchandise including Lego, "Disney 100 Wonders" Woolworths collector cards (see Figure 13), clothes, bags, crockery, homewares, and so much more.

Conclusion

In 1953, the Walt Disney corporation catapulted Tinker Bell into widespread popular culture and made her a cultural icon. In the original 1953 movie, Tinker Bell is portrayed as being sassy, possessive, jealous, and even spiteful; yet audiences continue to love her. Her portrayal earned her a spot as one of Disney's most loved and iconic characters, and Tinker Bell is now commonly referred to as the "unofficial mascot of The Walt Disney Company" (alongside Mickey and Minnie Mouse). Seventy years on from being 'adopted' by Disney, she has several feature films, Disney books, costumes, and accessories devoted to her.

The longevity of Tinker Bell's "zooming ball of light arcing over a silhouette of Sleeping Beauty's castle on the company's film division logo" very much demonstrates her role in Disney's corporate iconography (Davis 2008: 20–2). She "stands for the Disney enterprise: an animated logo, a dispenser of 'Disney Dust,' an official greeter at the gate and screen of the Magic Kingdom" (Bell 2016: 80). Despite the controversies her character has attracted over the years, Tinker Bell's magic *is* Disney's magic; or more to the point, it is Tinker Bell's magic that is marketed, commodified, technologised, and reproduced as saleable entertainment whether in theme park form or digestible media. As unofficial mascot, it is her magic that nevertheless facilitates the official tagline claiming the Disney parks to be "the happiest (or most magical) place[s] on earth".

Blair Speakman and Nancy Johnson-Hunt

True Blood (Alan Ball, 2008–2014)

Fairy, Fate, Agency, Hybridity, *True Blood*

Fairies abound in the popular imagination. In recent decades fairies have re-emerged on-screen, taking on forms that depart from traditional folkloric and literary expressions (Alexander 2014). In connecting fairy tales to our topic in question, Lorna Piatti-Farnell suggests that the "pervasive use of magical elements often coupled with the inclusion of monstrous creatures" is a distinct characteristic of the stories we have come to know and proliferate (2018: 95). Indeed, the HBO series *True Blood* (2008–2014), a page-to-screen adaptation from Charlaine Harris' literary *Southern Vampire Mysteries*, is in many ways an expansion of such fairy tale proportions. With regard to contemporary and traditional fairy tales, Piatti-Farnell admits, both fear and "conquering fear" with the anticipation of the hero's "ability to overcome adversities" is located firmly at the heart of the fairy tale (Piatti-Farnell 2018: 95). The televised *True Blood* series extends itself beyond even the recognisable parameters of the fairy in folk tale narration and models its appearance as the "not-so-distant cousin: "the popular culture vampire", (Koven 2012: 65). As far as fictional fairies go, however, *True Blood* follows the fairy-hybrid heroine Sookie Stackhouse (Anna Paquin) who is revealed to be on an undeclared quest of agency and liberation. Cast as the innocent and virginal 25-year-old, Sookie's hybrid identity is magnified through her relationships with Bon Temps' inhabitants, both vampiric and mortal. Her sense of agency and independence is captured adjacent to a dichotomy of co-dependence and her own self-doubts, which is made evident in her initial romantic courtship with vampiric character, Bill Compton (Stephen Moyer). Throughout the series, Sookie is endowed with a noticeable "proactive persona", a character who represents both a composed, enigmatic, and resourceful nature early on in the show and who goes on to question her fate on multiple occasions (Stasiewicz-Bieńkowska 2019: 233). Although Sookie has a desire to be "normal" and live a fully human life, her hybridity means she is frequently dehumanised and subsequently treated as an object by the vampires and supernatural creatures around her. In light of this, we endeavour to critically investigate Sookie's agency or lack thereof which connect her to her *own* fate rather than as an agent of fate to those around her.

Historically understood to be magical purveyors of fate, fairies and the fae occupy a place in the collective consciousness, navigating liminal planes that have much to do with notions of agency and control as they do with mortality and consciousness. As faeries are often "encountered on boundaries" they are rendered symbols for "other things" that cannot be acknowledged (Purkiss 2017: 83). It is precisely this ontological dubiety that this chapter will critically explore Sookie Stackhouse as an archetypal fairy figure and "virginal Southern belle" who traverses the liminal boundaries between mortal and celestial planes (Stasiewicz-Bieńkowska 2019: 232). In tracing the fate of Sookie more closely, our collective interest is concerned primarily in Sookie negotiating the terms of her hybrid identity. In doing so, we leave room to critically analyse more in-depth fairy tropes and gendered narratives for a future opportunity. We aim to analyse a series of sequential and noteworthy episodic examples across multiple seasons of *True Blood*. This chapter will therefore focus primarily on her relationships with vampire figure Bill and full-blooded fairy Claudine Crane (Lara Pulver). We will also aim to trace the ways in which her hybrid identity shapes her as a dichotomous figure, as she traverses notions of objectification and liberation.

Although the fae are not central characters in *True Blood*, they do however act as supporting roles to the adventures and trajectories of the vampires, making them significant figures in determining the fate of those around them. Sookie's significance to Bill is made manifest in *True Blood's* pilot episode entitled "Strange Love", when he visits Sookie in Merlotte's Bar, her place of work (Ball 2008). In the series, vampires can become addicted to fairy blood, so when Bill's attraction causes him to focus solely on Sookie, who is part fairy, has implications for her fate. Sookie's attraction to Bill is instantaneous, as she turns around to see him standing there waiting for her. Without so much as a word, Sookie is lured in by Bill's vampiric prowess. The most alluring quality, however, is that for the first time in her life, her telepathic abilities cease to function in his presence. Although she cannot hear his thoughts, her telepathy is overwhelmed by the thoughts of the surrounding patrons in Merlotte's Bar, who seemingly disapprove of this potential entanglement. Her path to Bill, although drowned out by the sound of the surrounding diners, forges ahead despite their reservations. As the episode comes to a close Bill asks Sookie: "What are you?" to which she responds, "I told you, I am a waitress". In a reflective gaze, Bill pronounces "No, you're something more than that. You're something more than human ... Sookie, that's an unusual name, Sookie". Finally, as the scene concludes, Bill stares intently at Sookie, as if others in the bar do not exist. At this moment, Sookie's fate is implicitly tethered to Bill as an embodied co-dependent union. The polarity between Sookie's innocence and lack of fear are to be a constant source of intrigue. When Bill asks her *what* she is, he further foreshadows what is to become of her identity. When Sookie is left perplexed by this interaction, it can be explained by the need to feel validated. As Eugene Doyen proposes, Sookie "cannot decide who she is" as her "social field defines her validated identity", however that her struggle "is to establish herself" within the social, cultural, and economic framework that surrounds her (2014: 48–9). How Sookie chooses to negotiate the terms of her existence, Doyen (2014) elaborates, is rooted in her habitus and, in essence, in her fate.

Throughout *True Blood*, the notion of fate is embedded within Sookie's fae identity. Indeed, this comes as no surprise as the English term fairy, or faery, as they are also known today, may have been derived from the Latin word *fatum* meaning fate, or *fata*, the goddess of fate (Knight 2002). However, Sookie's fate is usually controlled or at least partially governed by either Bill or Claudine. This can be witnessed in Season 3, Episode 9, titled "Everything is Broken" (Winant 2010), when Bill visits the fairy plane after consuming a copious amount of Sookie's blood. While in the fairy realm, Claudine is incensed, believing Bill "killed Sookie ... You have taken her blood. I can see it in you". Although Bill says he loves Sookie, Claudine rebuffs this claim: "You think you do. You only want her light ... Leave her alone vampire. We will protect her ... We have protected her". Claudine's confession that she and the other fairies have always protected Sookie is an example of how fairies are often depicted as other-worldly beings who visit earth with the intention of interacting with humans (Bane 2013). Although it is not clear who she is, or why she is concerned with protecting Sookie, it does appear that Claudine is concerned with vampires like Bill abusing or taking advantage of Sookie for her "light". Though tacitly encoded, Claudine's distress about Sookie's "light" suggests that she wants to shield Sookie from those who might want to use her fairy blood and powers for their own gain. In this way, being fairy – even being part fae – is framed as being highly desirable. The revelation that the fairies have "always protected her" suggests that Sookie has always lacked a sense of agency and free will.

Additionally, Episode 9 "Everything is Broken" (Winant 2010), is a pivotal narrative point within the series, as both Sookie and the audience begin to learn about the depth of Bill's manipulative behaviour. As Agnieska Stasiewicz-Bieńkowska (2017) argues, Sookie is frequently denied control over her body and her actions by her partners and remains in extremely exploitative and coercive romantic relationships. In the episode, it is revealed that Bill has been keeping a record of Sookie's background. Sookie confronts Bill: "Like that secret file you got on me … You know the one … Russel Eddington showed me … All kinds of birth records, family trees, newspaper clippings about me". Bill confesses he is investigating Sookie's life as there are many other vampires who appear interested in her for her telepathic abilities, including Bill's vampiric boss Eric Northman (Alexander Skarsgård). Sookie's hybrid identity, a source of fascination for Bill and the other vampires of Louisiana, mean they repeatedly invade her personal life and privacy in their quest to not only discover her fairy abilities. Moreover, it could be argued that Sookie's decorum and Southern charm means she is perceived as a diminutive figure by Bill, who frequently feels the urge to protect her. Bill digging into Sookie's family background, without her knowledge or consent, suggests he does not consider her own feelings or how she would feel about his investigation. In short, understanding Sookie's own feelings comes second to his desire to protect her.

As Season 3 progresses, Sookie must negotiate the tensions between her human–fairy status and the supernatural world. Sookie's revelation that Bill has been deceiving her throws her fate further into question. In "Everything is Broken", after confronting Bill for his secret investigation, he responds that he must: "know what you are, so I can protect you". However, Sookie continues to protest Bill's reticent and secretive behaviour, "you have to trust me, and stop thinking of me as a thing to be protected" (Winant 2010). During this scene, there are a few pregnant pauses, especially after Bill asks Sookie if she trusts him; Sookie's initial reluctance to answer his question heightens this moment of mistrust between them. Stasiewicz-Bieńkowska argues that the complexities in representing female agency, passion, and desire in *True Blood* "reveals an ongoing tension between patriarchal gendered expectations and an attempt to rescript female characters as agential sexual subjects" (2019: 238). This scene in particular reveals much about those ongoing tensions, as on one hand, Bill believes it is his duty to protect Sookie, including controlling her. In this instance, he withholds important information about Sookie's own identity and family history as well as his own actions which could impact Sookie's safety. On the other hand, this scene also highlights that Sookie does not accept Bill's protection and articulates her own frustrations that she is not trusted to protect herself – Sookie must fight for her own autonomy and sense of agency against her own partner.

Nearing the end of Season 3, in the finale episode titled "Evil is Going On" (Hemingway 2010), vampire figure Eric eventually reveals to her that Bill was directed by the vampire Queen of Louisiana, Sophie-Anne LeClerq (Evan Rachel Wood) to procure Sookie as an object, in the hopes of obtaining her fairy essence. Here as Sookie may think she has some form of control of her own life and actions, her life is filled with events she has little agency over. In this sense, Sookie's reactions to such events causes her to behave in a more dramatic fashion than her diminutive nature would suggest. Sookie's revelation that she is being investigated eventually fractures her trust in Bill, and she lambasts him: "You manipulated me into loving you ... love! You don't even get to use that word! I rescind my invitation". At this point Bill is hurled out of the door frame, which acts as a metaphorical portal between Bill's control over her, (re)assigning Sookie to steer her own fate. Sookie runs to the cemetery, a liminal space between those passed and still alive, and settles in front of her grandmother's grave, Adele Stackhouse (Lois Smith). The show's emphasis on graveyards being a meeting place suggests that fairies are "liminal in every possible way" sharing many characteristics of the dead, indeed they may also have the dead in them (Purkiss 2017: 84). When Sookie confesses that following her heart led her down a dead-end road, it is understandable that she is in the graveyard at this solitary moment. However, at this pivotal juncture, Claudine appears and gestures to Sookie that she is not alone placing her hand out to take it. Claudine implores Sookie to "Come with us", as she is joined by her fairy kin. In a true test of fate, Sookie stands up and walks over to her, as she places her hands outward a flash of light envelops the pair. This scene empowers Sookie's decision, that while she initially felt lost and alone, she is concurrently found and accompanied by her true kin.

In many ways, Sookie's hybridity means she is constantly being watched for either her own protection or otherwise hunted and procured. As Sookie navigates the disparate celestial and corporeal planes, she learns to tread lightly through the vampirically volatile Bon Temps district, but also the ethereal stratum of the fae. When Sookie's decision to accept Claudine's proposal takes her to the celestial plane of "Fairy", she discovers herself as the master of her own destiny. However, as quick as this decision is made, by the Season 4 premiere entitled *"She's Not There"* (Lehmann 2011), Sookie is back to being exploited as a commodity. For a brief period, she turns into an onlooker, watching her fate take an unusual turn in the hands of those around her. It is eventually revealed that upon her entrance into the fairy plane, Queen Mab (Rebecca Wisocky), a member of fairy royalty, sets out to harvest the "light" in Sookie as well as dozens of other human–fairy hybrids. In an ironic twist of fate, the life Sookie expected to eventuate from her leaving Bon Temps is thwarted at the hands of the fae who inhabit the celestial plane. As Doyen suggests, Sookie's need for solidarity, even with her own kin, becomes a strenuous task when social cohesion is already "strained and perhaps incompatible due to deep-seated and incompatible differences (2014: 23). Once protected by fairies from other supernatural creatures, she too becomes hunted by her own kin for their own nefarious reasons. Such incompatible differences between Sookie and her fairy kin drive her to return to Bon Temps. Despite Sookie's allyship and support for those around her, including her fellow fae, her social position "can be analogous to any individual wishing to promote or offer equality or fairness as a central tenet of the social order" (Doyen 2014: 23). While it has its disadvantages, ultimately Bon Temps is also a place in which Sookie remains proficient in navigating with more agency.

In tracing the origins of Sookie's fate, we have reflected on Sookie's character as "the plucky heroine, strongly principled but morally open-minded about others' misdeeds" (Mukherjea 2012: 110). This chapter raises important questions about long-standing discourse surrounding topics of female agency, objectification, and liberation. The notion that Sookie is a highly desirable plaything and a valuable possession not just for the vampires but also for the fairies further relegates her as a commodified figure. This commodification, although seen through the vampiric lens by means of their blood, is also called to attention through Sookie's narrative and the harvesting of fairy light. These very aspects of her own identity, her own physical fairy matter, are a constant point of contention for Sookie. Notwithstanding evidence to the contrary, Sookie is for all intents and purposes a fairy–human hybrid of magical and mortal proportions who may have more control over her fate than initially predicted. Her character delivers hope, that despite the adversities she experiences in both love and kinship, she is still a proponent of agency and liberation. In discussing Sookie's fate, we have reached an understanding that her decorous charm and erotic naiveté provides her with broad multi-dimensional appeal, which ultimately renders her the true master, or perhaps champion, of her destiny.

Amy Harris

The Daisy Chain (Aisling Walsh, 2008)

Introduction

Aisling Walsh, an Irish screenwriter and director, has been celebrated in Ireland for her exploration of rural Irish Catholic communities holding traditional beliefs about sex, religion, families, and women (Hergarty 2011). Walsh has received acclaim for being "one of few Irish female filmmakers [...] to make an impact on the horror genre" (Hergarty 2011).[1] Yet outside Ireland her work has been largely overlooked. An analysis of her first theatrically released and internationally distributed film *The Daisy Chain* (2008) will provide an opportunity to address this absence, consider her style, and examine how she negotiates cultural beliefs about motherhood through the language and iconography of Irish folklore and horror cinema.

The Daisy Chain explores the relationship between Martha (played by Samantha Morton) and an orphaned child Daisy (played by Mhairi Anderson), whose parents were killed in a house fire. The film situates traditional beliefs about fairy changelings within contemporary rural Ireland, a place where some parts of the community still hold strong superstitions about girls/women. The community believe that Daisy is responsible for the death of her family because a house fire took place "around Halloween when the veil between the living world and the afterlife is lifted", suggesting her supernatural affiliation. The community's claim might appear ridiculous to a sympathetic viewer, but

---

1   Since the release of *The Daisy Chain,* there has been an ostensible new wave of women working in horror in Ireland, including Aislinn Clarke, Lynne Davison, and Kate Dolan.

Daisy is cast out, unable to attend her local school after the villagers decide she must be a fairy changeling. *The Daisy Chain* explores the pervasiveness of these beliefs within the community and how the locals articulate their prejudice towards both Daisy and later towards Martha as she assumes a parental role over the child. Through the complex relationship of a mother and a fairy child, *The Daisy Chain* offers an especially valuable example of maternal representations in Irish film. Walsh uses tropes and themes linked to culturally resonant folklore to push back against damaging beliefs about motherhood, and to articulate women's experiences of pregnancy, birth, postpartum trauma, and maternal desire.

Fairy Changeling Lore

In folklore, an Irish fairy changeling is an unbaptised fairy left in place of a human child. Some Irish folklore tales mark physical or mental differences in children as evidence of a changeling. For example, typically, a changeling is described in European lore as sickly, often unable to develop speech or grow at the usual rate of a child (Ballard 2014). Superstitions about changelings are evidently a more palatable way of normalising ableist discourse facilitating prejudice and violence (Eberly 1988). Nevertheless, compounding myth with social prejudice, it has been noted that:

> from pre-Christian until recent times, many people have sincerely and actively believed that supernatural beings can and do exchange their own inferior offspring for human children, making such trades either [to] breed new strength and vitality into their own diminutive races or simply to plague humankind (Ballard 2014: 137–8).

Although the dominant themes in this lore are "anxiety, guilt, and fear [about] neonatal deaths, infanticide, congenital defects, and unexplained illnesses in early times [...] familiarity does not seem to have blunted the associated emotions [towards changelings]" (Simpson 2000: 27). In fact, to this day if a child is born with any health conditions, they are commonly referred to as "fairy-struck" (*Irish Post* 2021). This trope is explored in *The*

The Daisy Chain (Aisling Walsh, 2008)

Daisy Chain where a disturbed child, Daisy, is assumed to be a changeling by the superstitious villagers. Walsh explores some of the negative reactions towards the fairy changeling to unpack trauma around children, families, and motherhood at local and wider cultural level.

It should be noted that although the film addresses the connection between disabled children and changelings, it does not resolve the crisis of representation. Throughout the film Daisy's onscreen presence becomes increasingly unsettling. In a notable sequence, the camera tracks Daisy sneaking into the hospital to play with a newborn baby. Daisy looms over the hospital cot and the camera captures her sinister smile from a disempowering, low angle, teasing the viewer with notions of her immorality. To re-emphasise association with superstition, the film ends with a closeup shot of Daisy's sneering face, coupled with the ominous soundtrack of a child's nursery rhyme. These stylistic choices, typically associated with suspense building in the horror genre,[2] suggest to the viewer that perhaps Daisy was evil all along, entrenching the ideas the film is, on the surface, engaging with. Although this problematises the film from that perspective, as the following analysis reveals, Walsh nevertheless engages with Irish folklore's patriarchal effects which frame girls/women as untrustworthy, mischievous, and dangerous, reinforcing their domestic confinement and presumed need to be controlled (Brophy 2022).

Film Analysis

The plot follows couple Martha (a Londoner), and Tomas (an Irish man, played by Steven Mackintosh), who move to a remote village on the west coast of Ireland to set up a home and work through the premature death of their firstborn daughter, Chloe. The film spends time introducing the rural community, situating the narrative within a specific local context. This is important because Walsh is critiquing a very specific set of beliefs held within

---

2   Although not the focus of this chapter, Walsh' choice to draw on the stylistic conventions of horror children is problematic and would be interesting to explore further.

the imagined community. However, the film could be accused of reinforcing stereotypes of soft primitivism in its allusions to changelings and simple country folk. It is worth noting that Walsh is from Dublin which suggests that there is an implicit rural othering throughout the film. To transport the viewer into this community, the film opens with the couple driving back to rural Ireland to take up residence in Tomas' former childhood home. The radio plays in the car and the presenter hosts in Gaelic. Martha mocks Tomas' familiarity with the language, asking "is this what you used to listen to", indicating her less remote origins and foreshadowing the cultural clash that will ensue. The muted colour palette has a flattening effect which foreshadows a claustrophobic, insidious return to tradition.

The couple arrive at a dilapidated farmhouse where Tomas excitedly promises a disappointed Martha that he will rebuild the old family home. Martha is pregnant and worries that the space will not be ready in time for the birth. Her anxiety regarding pregnancy and childbirth are clearly exacerbated by the traumatic death of their firstborn. Martha's longing for motherhood conflicts with her difficulty in having a child of her own. Her representation as a maternally desiring figure without a child of her own marks her as an outsider in the context of traditional superstitions. When Martha first arrives, every woman she interacts with has a child, marking Martha as removed from the other women. The suggestion that motherhood is a natural state for women living in the rural village highlights the return to a patriarchal imaginary among the community, implying that women must fulfil traditional roles as mothers and homemakers.

Shortly after arriving, the couple meet Daisy, who is autistic and, in line with folkloric beliefs about ostensibly "different" children, believed to be a fairy changeling by the villagers. Daisy embodies common traits that can be identified throughout changeling lore. She is very small, unkempt, has unusually dark eyes, and a fussy temperament, which corresponds with common motifs regarding changelings (Thompson 1958). Convinced that the rumours are nothing but superstition and appalled by her neglect and poor treatment by the villagers, Martha decides to care for Daisy.

Daisy responds positively to Martha's affections and their bond only strengthens the locals' beliefs that there is something wrong with Martha too. They warn Martha that Daisy is a mischievous and evil fairy who will corrupt and put Martha and her unborn baby at risk. Appalled by this, Martha is sympathetic towards Daisy and assumes a maternal role. As the emotional bond between Martha and Daisy develops, Martha and Tomas' relationship becomes strained, prompting him to confront her about the loss of their daughter and express concerns that their newborn child, whom they know will be a boy, will be overshadowed by Martha's preoccupation with Daisy. Martha finds the beliefs regarding Daisy disturbing and begins to research autism in children to try and understand her needs and support her. Martha, culturally removed from the superstitious beliefs, understands that Daisy has particular behavioural patterns: intense emotional outbursts; a limited vocabulary; a fear of physical contact; a tendency to isolate herself from others; and a habit for anxious, co-dependent attachment. Martha therefore poses a threat to the community for reminding them of their prejudice towards seemingly "different" children.

Once both are marked as outsiders by the community, Martha and Daisy begin to mirror each other.[3] This is made explicit in a scene where Daisy, with Martha's help, dresses up as her surrogate mother, stuffing pillows in her dungarees to emulate a pregnant stomach.[4] When Tomas discovers the pair, he is horrified and reminds Martha that she should be thinking about their own child instead of Daisy. Tomas' reminder to Martha is, on one hand, a warning about the intensity of her relationship with Daisy and, on the other, an accusation that she should focus her energies on her own child, although there is no suggestion of what this looks like.

The unpleasant implication here being that an adopted child cannot be loved the same as a biological one. This further reflects societal expectations about the structure of the family unit, which dictates that having an adopted child is not the accepted norm in this community, and being an adoptive mother is not the same as being a mother to one's own child. Nevertheless, Martha's blossoming relationship with Daisy is clearly helping

---

3   Throughout the film physical space between Martha and the community starkly visualises her separation from them. For example, at the school gates the mothers stand close together and away from Martha, who is usually stood alone or only with Daisy, symbolising her otherness as the sole sympathiser to Daisy's needs.

4   Although this chapter does not have the scope to explore the dress-up scene further, it also points to a process of "girling" where Martha is "playing" at being a mother to Daisy.

her process her own grief about the loss of Chloe, indicated by the freedom she finds in caring for Daisy as they playfully do crafts together – which reignites Martha's passion for painting – go on long walks, and read stories. However, Daisy, traumatised by her experience of neglect and abuse, forms an anxious attachment to Martha smothering her in excessive attempts for constant affection. Daisy could be described as a challenging child due to the demands she places on Martha for attention, affection, and reassurance. In her desperation to become a mother, Martha is delighted when Daisy calls her "Mummy" and decides to overlook Daisy's behaviours. In many ways, Daisy can be seen as a vehicle to invite conversation for the viewer about the challenges of mothering (although this could make a patronising suggestion about Martha's lack of preparedness for the varying challenges that come with motherhood), and the emotional labour involved in parenting alone. Trapped with each other's vulnerabilities, an increasingly stressed Martha begins experiencing complications with her pregnancy, which reinforces the community's belief that Daisy is the source of Martha's ill health. This is due to suspicions about Daisy's supernatural abilities, rather than recognising that Martha is under a considerable amount of stress and still grieving her loss.

Martha's grief is always present, although perhaps most heavily symbolised through the muted colour palette of the landscapes. When Martha arrives in the village, the sparse and bleak scenery reflects her depressed emotional state. It is only when Martha bonds with Daisy that these landscapes are softened by golden hues of sunshine and clear blue skies, a ray of hope reflecting Martha's delight at finally fulfilling a maternal role. Overjoyed by her relationship with Daisy, Martha begins the fostering process, although she problematically does not consult Tomas. The lack of communication between Martha and Tomas is, in part, due to Tomas' traditional understanding of familial structures inherited from the community. Whenever Tomas encourages discussion about social care for Daisy, Martha makes a point of going it alone. When Tomas tries to talk to Martha about their baby, Martha is dismissive. In this way, Tomas' increasing frustration with Daisy could also be read as a frustration towards Martha's independence and her unwillingness to think about her prescribed role as a (natural) mother in their growing family – a family he can only imagine within a traditional framework. However, when strange occurrences start to happen, the landscapes increasingly reflect Martha's fraught temperament and fear that Daisy will be taken away, removing Martha's burgeoning identity as Daisy's mother. Images of crashing sea waves are intercut with thunderous landscapes shot in the dark of the evening, as Martha's grief and distress return.[5] In one such scene, Martha visits Tomas' friend in hospital after she has given birth to a baby boy. The friend advises Martha that when she has her own child, she should ask for "as many drugs as you can get" to soften the pain of childbirth. Although the comment is made in jest, the viewer knows that Martha has already experienced labour and can understand why Martha winces at the joke. The proceeding shots of the hospital return to a grey and muted colour palette, echoing the return of Martha's sadness.

In the final scenes of the film a garda reveals to the increasingly distressed and exhausted couple that Daisy tried to smother her younger brother before leading him to the sea where he drowned. Tomas demands that Martha choose between him and Daisy, reiterating that she is not Daisy's mother and should focus on mothering their own child. It cannot be denied that Martha experiences pressure from the community and Tomas to fulfil the expectations of traditional motherhood. Martha's representation as maternal figure without her own child highlights the return to a traditionalist imaginary (Collins 2020). Walsh questions these traditional beliefs by situating a modern woman from an urban space in an Irish rural community that upholds traditional family values. The cultural clash sees the protagonist focus on her maternal identity above anything else, even at the expense of her marriage to Tomas. Martha refuses most of the social care support, performing the role of mother, carer, therapist, and teacher to Daisy; and wife and homemaker to Tomas. Her maternal identity becomes all-consuming. Martha succumbs to the pressure to fulfil the unobtainable goal of the perfect mother, encouraging the viewer to question the idealised self/mother. Martha's desperation to care for Daisy eventually results in the collapse of her marriage and, overhearing this, Daisy becomes hysterical. When Martha gently questions the child about the death of her brother, Daisy presses on Martha's stomach, possibly attempting to hug

---

5   Pathetic fallacy is also indicative of Martha's alienation and the community's hostility to outsiders.

her, before pushing her over. At the end of the film Tomas returns to find a bloodied Martha collapsed on a soaked bathroom floor; meanwhile, Daisy is in the adjoining room swaddling the couple's newborn baby. It is unclear whether Martha survives the traumatic birth, but Tomas takes the newborn away, leaving Daisy and Martha alone.

Conclusion

The Daisy Chain is a valuable example of pushing back against limiting and troublesome representations of gender in rural Ireland, through an exploration of fairy changeling folklore and its relationship to myths about motherhood. The ambiguity surrounding Daisy's fairy identity could have been better executed to underscore some of the film's clear aspirations; however, by exploring struggles around motherhood and societal expectations of maternity in rural Ireland, the film offers a complex negotiation of what that label confers (Pisters 2020). With matters around pregnancy, childbearing, and parenthood bearing often not only on women's bodies, but also on the bodies of some non-binary people and trans men, many of these themes have been taken up by marginalised filmmakers, although they have not been widely acknowledged, at least until recently. Walsh brings together opposing discourses about fairies and motherhood to explore how these topics impact each other, wading through damaging superstitions about girls/women in rural Ireland to offer a portrait of the impact they have on the women forced to bear the burden.

Throughout the film there is the sense that pregnancy, birth, and motherhood is complex and not wholly positive, but also a source of anxiety, anger, depression, and an identity that can be consuming (Fischer 2016).[6] For Martha, the challenges also lie in the fact that she is mothering alone because

---

6   Almost literally if we consider the bodily invasion and subsequent growth of a baby associated with the pregnant body.

The Daisy Chain (Aisling Walsh, 2008)

Tomas becomes more distant and withdrawn, even seeking out an affair with an old flame when he feels that Martha's attention is too focused on Daisy. Ultimately, *The Daisy Chain* offers a complex and sensitive portrayal of motherhood as one that is equally as enjoyable as it is fraught, but also depressively doomed.

Rebecca Wynne-Walsh

A Cinderella Story (Mark Rosman, 2004)

Fairy tales act as "cultural barometers", allowing audiences to critically question the "values, traits, and ideologies of the tale's world" and, by extension the world we ourselves live in (Ju Lee 2022: 37). The story of Cinderella is undoubtedly one of the most well-known fairy tales across global cultures. The story features many iconic elements, not least of which is the Deus ex machina figure in the form of the fairy godmother. Although each version presents a unique configuration, the fairy godmother consistently appears in some form to aid the heroine in her struggles. This chapter explores, indeed problematises, a conceptualisation of the fairy godmother as a catalyst for the post-feminist neo-liberal self-determination of the heroine. During this process the identity of the fairy godmother is marginalised. The fairy godmother's happiness and success are entirely entangled with those of the heroine. The figure demands critical reassessment in intersectional terms. Here, her function is addressed in relation to class, gender and racial power dynamics in the Cinderella-fairy godmother relationship in the hopes that a more critically nuanced version of the character may be imagined. This contributes to an endeavour towards the granting of dreams and wishes that recognise diverse and inclusive socio-cultural values.

The need to strategically reconstruct representations of this character is vital in 21st century media which increasingly prioritise diversity, inclusivity and intersectionality. By way of illustrating this, I analyse the fairy godmother role occupied by Rhonda (Regina King) in *A Cinderella Story* (Rosman 2004) in terms of her authorial and maternal function. Although *A Cinderella Story* is a version inflected by the neo-liberal, post–*Sex and the City* (1998–2004) popular culture climate of the early 2000s, the core premise remains much the same. Upon the sudden death of her father, a young girl falls victim to poverty and domestic enslavement at the hands of her cruel stepmother and her equally cruel daughters. The heroine's inherent beauty, both inside and out, gains the attention of fairy godmother figure who transforms Cinderella's clothing and grants her the ability to attend the royal ball and gain the prince's favour. The prince naturally falls in love with Cinderella and their marriage ensures our heroines rapid rise from peasantry to princesshood. While various adaptations of this narrative make certain adjustments and omissions, a fairy godmother, in one form or another, remains consistently present. Kay Young insightfully describes the fairy godmother as an "authoring" figure (2011: 288). The fairy godmother actively facilitates Cinderella's makeover into the belle of the ball and eventual princess. This chapter questions Rhonda's actual narrative authority and agency in her fulfilment of this authoring role.

The reconsideration of the fairy godmother in this cult classic complicates understandings of the figure in relation to "power, class, gender, and social norms" with specific focus on Rhonda's service industry employment as diner manager and the complexities of her relationship with Sam (Hilary Duff), the film's Cinderella character (Ju Lee 2022: 37). Gomes highlights that hospitality and service industries represent "stereotypes, assumptions, communicative rationality and instrumentality" (2023: 2). For this reassessment of the fairy godmother to offer an intersectional consideration of the figure in terms of class, gender, and race, then the recurrent placement of the character in servient positions must be central.

As the majority of Rhonda and Sam's interactions take place during their shifts working at the diner, each characters' values and responsibilities are performed through the hospitality industry context (see Figure 14). The "intersection of hospitality and popular culture" as witnessed in *A Cinderella Story* presents a "dynamic space where societal values, norms, and human interactions are vividly depicted and often critiqued" (Reymond 2023: 1). This setting allows for a performance of "society's ever-evolving ethos" surrounding both feminism and femininity (Reymond 2023: 1.). If neo-feminism operates at the intersection of capitalism and feminism, then hospitality operates at the intersection of interpersonal relationships and capitalism. Examining *A Cinderella Story*, and the fairy godmother figure, in this context forces the viewer to challenge their relationship with intersectional feminism when presented with culturally

A Cinderella Story (Mark Rosman, 2004)

Figure 14. Rhonda (Regina King) and Sam (Hilary Duff) dealing with customers at the diner. *A Cinderella Story,* directed by Mark Rosman (Warner Bros. Pictures 2004).

loaded images of racial diversity, social class, employment status, marital status, and gender roles.

A Cinderella Story updates the classic tale as a star-vehicle for, then, teen idol Hilary Duff. Duff occupies the Cinderella role as Samantha "Sam", a young girl whose kind father's unexpected passing leaves her in the "care" of her selfish stepmother and stepsisters. Sam is forced to keep house for the women whilst working in her father's diner which stepmother Fiona (Jennifer Coolidge) has turned into a tacky and impractical establishment. Rhonda now manages this diner in an effort to protect part of Sam's father's legacy. In between intense school studies, Sam embarks on a romantic relationship online with an unknown boy from her school with whom she bonds over shared desires to attend Princeton. It is worth noting that, in this text, Sam's love interest, her "prince", operates more as a symbol for her financial and academic success rather than the more traditional fairy tale marriage plot. Sam and her secret love agree to meet at the masquerade ball being thrown at their high school. Sam's plans are thwarted by Fiona's demands that she works an extra shift in the diner. However, Sam's coworkers are determined to save the day and diner manger Rhonda takes it upon herself to get Sam to the ball and find her "bliss".

The bliss Sam strives to achieve is tied to the value this film imposes on a "neoliberal feminine identity" in which femininity, and indeed feminism, is inextricably linked to capitalist consumer culture (Kennedy 2018: 429). The heroine's success is an independent journey of self-determination and individuation achieved through practices of consumption, financial success, and upward social mobility. However, Sam's success comes at the expense of her "fairy godmother" Rhonda's diligence, loyalty, and personal and professional sacrifices.

The hospitality industry setting of *A Cinderella Story* literalises the service roles occupied by both Cinderella and her fairy godmother in all traditional tellings of this tale. When Sam's hopes of attending the climactic ball are dashed, Rhonda saves the day. In Sam's hour of need, Rhonda takes advantage of the respect her managerial position affords her among local hospitality and retail workers to get Sam's shift covered and to engage in some after-hours costume shopping. Several films which deploy and/or modernise the Cinderella/Pygmalion narrative include characters who are employed within the hospitality industry or distinctly servient professions. While this chapter addresses Rhonda specifically, other obvious examples of this trope include *Pretty Woman's* (Garry Marshall 1990) hotel manager Barnard Thompson (Héctor Elizondo). Elizondo's character plays a vital role in showing Julia Robert's protagonist prostitute where she may shop for fancy clothes to redefine her appearance and how she may conduct herself in altering her table manners so that she may be accepted in high society (see Figure 15). In *Ella Enchanted* (Tommy O'Haver 2004), a jukebox musical reimaging of the Cinderella narrative, the fairy godmother figure is a "house fairy" (Minnie Driver), a character coded as a maid in Cinderella's family home. Another key example features Elizondo again; this time as the loyal bodyguard Joe who protects and chauffeurs a young Mia (Anne Hathaway) as she embarks on her transformation from shy high school nerd to glamorous princess in *The Princess Diaries* (Garry Marshall 2001).

Hilary Radner notes that neo-feminist film narratives – such as *A Cinderella Story* – are defined by a glorification of the "rewards of work and career" while romance and the marriage plot "play a secondary role" (2011: 102).

A *Cinderella Story* (Mark Rosman, 2004)

Figure 15. Hotel manager Barnard (Héctor Elizondo) teaches prostitute Vivian (Julia Roberts) how to use her cutlery in a high society context. *Pretty Woman*, directed by Garry Marshall (Touchstone Pictures 1990).

Rhonda and Sam's relationship, as fairy godmother and Cinderella respectively, conforms to Radner's definition. If the fairy godmother figure facilitates the fulfilment of the heroine's dreams, then it is significant in this neo-feminist context that the "bliss" Sam finds in this film is her acceptance to Princeton University. In this text, Rhonda also finds a happy ending, in becoming part owner and manager of the diner. However, if we consider the fairy godmother as a catalyst for post-feminist neo-liberalist ideals of self-determination, then the service industry context alters the power dynamics at play in this narrative. These happy endings illustrate the divergent feminisms represented by each character. Sam's happy ending conforms to the neo-feminist dream of self-determination and personal fulfilment. In contrast, Rhonda's happy ending can be read as a success in line with the values of second-wave feminism tied to radical intersectional equality, resolving class conflicts and engendering interpersonal support. Radner posits, "While second-wave feminism did advocate a program of self-fulfilment, it did so within a climate of social responsibility and state intervention. Individual fulfilment was meaningless outside the policy of larger social and institutional change" (2011: 9). Radner's distinction between second wave and neo-feminism is performed in *A Cinderella Story* as Rhonda's happy ending is conditional on Sam's. Rhonda can only find her own "bliss" if she first enables Sam to find hers. This limits the feminist value of Rhonda's professional success. This achievement is bound to Sam as the heroine herself is Rhonda's co-owner. As Sam could not achieve her success without Rhonda, the fairy godmother figure occupies a complicated position. It is Rhonda's knowledge, courage, job, and indeed possessions that allow Sam to fulfil her dreams. This supports the aforementioned "authoring function" of the fairy godmother while undermining her independent narrative agency.

Ju Lee asserts social class is a crucial factor in fairy tale narratives such as this (2022: 45). While Sam works at the diner, the audience is privy to the knowledge that the diner is in fact her middle-class birth right via her father. Rhonda does not share Sam's family heritage and so her story is marginalised reflective of her lower-class status in the diegesis. Although the tale begins with Sam as a waitress and exploited teen worker, it is Rhonda who is forced to remain in a servient position in the diner after facilitating Sam's self-discovery journey, culminating in her Princeton acceptance. While the fairy godmother has access to narrative authorship, she is denied access to the associated authority. Rhonda is instrumental in Sam's neo-liberal self-determination, but this function comes at her own expense. In this sense, an intersectional re-reading of the power dynamics and narrative authorial tensions between the Cinderella figure and her fairy godmother allows us to critically address such "deeply embedded ideologies related to power, class, gender, and social norms" (Ju Lee 2022: 45).

In the opening of *A Cinderella Story* not only is Rhonda manging the diner where Sam is forced to be a waitress, Rhonda is the only positive female figure in Sam's life. As articulated by Klenk, in all versions of the Cinderella tale, our heroine has lost her mother, a trauma which is "compounded by the speedy introduction of her father's second wife with her two children" followed by the loss of her father (2023: 80). Rhonda's leadership position, kind demeanour and vocal dislike of Sam's stepmother contribute to understanding the fairy godmother as a maternal substitute. The case of Rhonda and Sam performs an age-old stereotype of maternal sacrifice in favour of the

A Cinderella Story (Mark Rosman, 2004)

daughter's success and upward social mobility. The fairy godmother figure is coded as not having access to nor desire for the same opportunities the heroine seeks. While Fiona forces Sam to work in the diner to pay for university tuition, this establishment is Rhonda's entire professional pathway. For Sam, the service position is a punishment; for Rhonda it is vocational and maternal. At one point, Rhonda calls attention to the fact that she could work in any diner but chooses to stay here and look after Sam in loco perentis. Rhonda as fairy godmother thus offers the intimacy, recognition, assurance, and comfort traditionally provided by a mother figure (Klenck 2023: 80). This is not the only instance of Rhonda sacrificing her desires to facilitate Sam's. When it is decided that Rhonda and her co-workers will aid Sam in attending the ball, Rhonda offers Sam her own wedding dress. The repurposing of Rhonda's dress symbolically replaces her hopes with Sam's, devaluing Rhonda's aspirations in favour of achieving Sam's.

Such plot points exemplify the core function of servient fairy godmother who enables Cinderella's social mobility while sacrificing her own. This is a widely occurring stereotype in representations of maternity onscreen. As articulated in quite literal, if problematic, terms in Greta Gerwig's instantly iconic neo-feminist text *Barbie* (2023), "we mothers stand still so our daughters can look back to see how far they've come". Rhonda's maternal function further calls attention to the gendered tensions between heroine and fairy godmother. Sam and Rhonda's affectionate mother–daughteresque relationship, and their consistent location in the diner which ties them, particularly Rhonda as Sam leaves for university, to feminine fairy-tale character traits of domesticity, and nurturing (England, Descartes and Collier-Meek 2011: 559).

The reconsideration of the power dynamics between heroine and fairy godmother in gendered and class-based terms allows for a strategic reconstruction of this enduring fairy tale in recognition of 21st-century values of quality, diversity, and inclusion. As per this impetus of this reconsideration of the fairy godmother figure, each representation of femininity must be addressed in intersectional terms. In her fairy godmother role, Rhonda fulfils the maternal function in Sam's narrative. This ascribes the responsibility for the success and well-being of the young white heroine on the shoulders of a working-class Black woman. As such the racial dynamics of this Cinderella–fairy godmother relationship is reflective of the centuries-long tradition within the United States, particularly of "Black women taking care of white children as 'the help'" (Galvan 2022). Several contemporary retellings of the Cinderella tale on screen present a fairy godmother figure who is racially distinct from their respective Cinderella. The white Vivian (Julia Roberts) is paired with the supportive and protective Barnard played by Puerto Rican actor Héctor Elizondo. The 2021 Amazon Prime Original musical version of *Cinderella* (Kay Cannon) casts Black queer icon Billy Porter as the Fabulous Godmother alongside the visually racially ambiguous Camila Caballo as the titular heroine (see Figure 16).

In some ways the narrative authority of the fairy godmother figure has evolved positively. Rhonda is a Black hospitality worker who must sacrifice her happiness to enable the success of the white heroine. Her actions define the entire unfolding of Sam's self-determination journey, yet Rhonda's bliss is strictly contingent on Sam's happy ending. If we contrast this with Billy Porter's Fabulous Godmother, Fab G as they are referenced in the film, we can

Figure 16. Billy Porter as Fab G, arrives to send Cinderella (Camilla Caballo) to the ball. *Cinderella*, directed by Kay Cannon (Amazon Prime 2021).

see some positive developments. First and foremost, the authoring function of the fairy godmother is signalled from the outset of this version as the opening sequence is narrated by the instantly recognisable voice of Porter. This places a queer Black voice in the position of voice-of-God narrator. Furthermore, Cinderella's success in the 2021 version is a far more collaborative process. While Rhonda presents Sam with her own wedding dress as a ball gown, Fab G magically conjures Cinderella's gown directly from one of the heroine designs for her planned dress-making business. While Sam's servient work in the diner is merely an avenue towards perceived upward social mobility in university; 2021's version of Cinderella dreams of her own business developed from the skills she has learned through her service position. While Rhonda sacrifices professional mobility out of duty to Sam, Fab G chooses to aid Cinderella because she freed him from the cocoon which had entrapped his caterpillar form. Both texts put a premium on hard work but there is a distinct power imbalance in *A Cinderella Story* which valorises Sam's work but positions Rhonda's work as a means to Sam's end. In 2021's *Cinderella,* while the fairy godmother figure occupies less screen time, Fab G makes a choice to aid Cinderella based on a mutually beneficial relationship thereby positively realigning the existing power dynamics between the characters. A re-evaluation of the narrative function of beloved fairy-tale characters concomitant with changing contemporary representations offers audiences a lens through which gender, race, and class in these texts may be understood. This in turn allows for more sophisticated critical nuance in reimaginings of the figure so that they may grant wishes in a fairy tale with intersectionally inclusive values.

Part IV

By Any Other Name

Quink

Kirstin A. Mills

Jo Anna Burn

The Elves (Terry Pratchett, 1992–2003)

Evil, Free Will, and Redemption

> Elves are wonderful. They provoke wonder.
>
> Elves are marvellous. They cause marvels.
>
> Elves are fantastic. They create fantasies.
>
> Elves are glamorous. They project glamour.
>
> Elves are enchanting. They weave enchantment.
>
> Elves are terrific. They beget terror.
>
> The thing about words is that meanings can twist just like a snake, and if you want to find snakes look for them behind words that have changed their meaning.
>
> No one ever said elves are nice. Elves are bad. (Pratchett 1992)

Sir Terry Pratchett (1948–2015) was a prolific writer and is best known for his humorous Discworld fantasy novels, but he was also an extremely well-read folklorist. Although he modestly writes, "I did not study folklore any more than a butterfly studies flowers" (2000: 8) his work is grounded in a deep understanding of folklore to the extent that in 2008 he co-wrote with Jacqueline Simpson *The Folklore of Discworld*. Pratchett's bestiary is very rich and features many magical creatures such as banshees, dwarves, trolls, and werewolves who are derived from ancient folk tales. However, the word *fairy* is used infrequently and mainly to describe the Tooth Fairy which is an anthropomorphic personification that began life as the personification of fear as the first Bogeyman, but later evolved into a (more or less) benign being that protects children from sympathetic magic by safeguarding their teeth (*Hogfather* 1996). In typical whimsical Pratchett fashion, the position of tooth fairy later becomes a franchise with individual species developing their own unique brand of tooth fairy to suit the needs of their people. The first troll tooth fairy Clinkerbelle was arrested by the Watch on suspicion of being an imposter who was only interested in profiting from troll children's diamond teeth. Fairy Godmothers also feature on the Disc, notably in *Witches Aboard,* but they are not always the benign beings of Disney and are capable of acts of magical terror once they "begin to cackle" as Esme Weatherwax puts it.

Most fairies on the Disc are not given the title fairy, even though they may have been inhabitants of Fairyland. Fairyland is presided over by Nightshade, the Queen of the Elves, and it is elves in Discworld who are the true inheritors of the fairy mantle from Round World (Earth) folklore. Terry Pratchett was very interested in the folklore of many cultures, particularly Scottish, Irish, Welsh, and English folk beliefs. He describes it as providing "depth" to society (2000) and the folklore of the Disc is firmly entrenched in the folklore of Round World. In Celtic folk lore fairies or fae can inhabit a range of behaviours from merely annoying for example by playing pranks such as causing milk to sour or a horse to go lame, to terrifyingly malevolent, such as stealing children and replacing them with empty changelings. In Celtic mythology and on Discworld, elves are not benign.

Other species of fairy in Discworld include the Nac Mac Feegle (also known as Pictsies), who left Fairyland after a dispute with the Elf Queen (*The Wee Free Folk* 2003), and the gnomes and goblins who are essentially moral and benign creatures, in character free from elf malevolence. The Nac Mac Feegle are some of Pratchett's most loved characters who remain brave and loyal to their "Big wee hag" Tiffany Aching, despite their tendencies towards drinking, fighting, swearing, and thieving.

According to the folklorist, and Pratchett consultant Jacqueline Simpson (2011: 77),

> "Elf" and its cognates are ... native terms in several Germanic languages for minor supernatural beings, as opposed to gods; they are derived from an Indo-European root meaning "white." "Elf" was a standard word in English until it was gradually superseded, from the thirteenth century onwards, by the more elegant and aristocratic French loanwords "fairy" and "fay".

Therefore, elves and fairy are essentially the same and refer to the fair faced and attractive, but morally dubious species. Pratchett names the elf lords Peaseblossom and Mustardseed after the fairies in *A Midsummer Night's Dream*, but beyond the names, Pratchett's elves retain very little in common with Shakespeare's characters. Simpson (2011) describes elves as morally ambiguous, not entirely good, and not entirely bad, but for Pratchett few beings in the Disc cycle are as unrelentingly unpleasant as elves. They are the oppressors of all other beings, totally lacking in empathy. Granny Weatherwax describes them as "cruel for fun, and they can't understand things like mercy. They can't understand that anything apart from themselves might have feelings" (Pratchett 1992: 157).

Pratchett likens them to cats in their beauty, self-absorption, and casual acts of cruelty. This dark view of the Elven race has its origins in Round World folk history and is echoed in Pratchett's choice of name for the Fairy Queen's henchman Lankin. According to the historian Graham Seal (2018) the ballad of Long Lankin first appeared in print in 1775 published by the British ballad collector Bishop Thomas Percy, although it is certain that the text is much older. It is a nightmare tale of the torture and murder of a baby and its mother by the bog-dwelling Lankin who exacts terrible revenge for being cheated by the lady's husband (see Figure 17). The elves of Pratchett cleave to this dark history, unlike the rehabilitated elves of Tolkien who retain only vestigial cruelty such as the Wood Elves' treatment of Thorin and company in *The Hobbit*.

In *Lords and Ladies* Granny Weatherwax describes the elves thus: "When they get into a world, everyone else is on the bottom. Slaves. Worse than slaves. Worse than animals, even. They take what they want, and they want everything" (1992: 156). Although she does admit that they are beautiful and stylish. However, the glamour that elves exert over non-elvish races only persists whilst the elf has power and is conscious. When unconscious they resemble "a long, thin human with a foxy face" (Weatherwax 1992: 155).

On the Disc elves are almost universally evil; they live in a parasitic, snow-bound world and abduct creatures from outside their realm holding them prisoner and tormenting them pitilessly. They are immortal and able to read people's minds and tamper with memory, blinding them with glamour. Elves can be seen as Pratchett's depiction of humanity at its worst; taking pleasure

Figure 17. The image is by Paul Kidby from the *Folklore of Discworld* (2000). Reproduced with the permission of the artist.

from the suffering of others, elves demonstrate a depraved morality that is unrelieved by any sense of pity. However, things were not always thus. In *The Wee Free Men* we learn that Fairyland used to be a warm living place of Summertime, but a rift between the Elf King and Queen led to the King moving to his own realm in an underground barrow, The Long Man in Lancre. This place is as uncomfortably warm as the Queen's realm is frigid. Thus, Pratchett uses the elves to discuss one of his favourite themes: gender politics.

Gender roles are a key theme in Pratchett's work (Sinclair 2015) and he loved to challenge accepted norms. As Lymbou (2015) points out, witches and wizards are both powerful magic practitioners, but the witches employ a more practical form of magic. On the Disc witches deal with dirty jobs and serve the community as midwives, carers, and layers-out of the dead. Wizards tend to eat large meals and generally avoid doing any real work. It is clear which side of the gender battle Pratchett's sympathies lie. In 1985 he said in a talk *Why Gandalf never married*, "The fantasy world in fact, is overdue for a visit from the Equal Opportunities people because, in the fantasy world, magic done by women is usually of poor quality, third-rate, negative stuff, while the wizards are usually cerebral, clever, powerful, and wise". Pratchett set about rectifying the sexism in the fantasy genre and challenging other tropes and norms. As Sinclair (2015) remarks, "Pratchett is a master of engaging with the conventions of fantasy in order to subvert them along with our own imaginations about historical truth" (17). Many of his female protagonists are forced to fight for a more equal placement in the world, such as Eskarina Smith's struggle to become the first female wizard in *Equal Rites* (1987), or the right to freely express themselves as women in male societies such as the dwarf Watchman Cheery (Cheri) Littlebottom in the *Watch* novels (1989–2011), and the golem Gladys in *Going Postal* (2004), and *Making Money* (2007). Previously women either had to make the best of being second class in a male-dominated world or pretend to be men like Sergeant Jackrum and fellow enlisted soldiers in *Monstrous Regiment*. In a further act of defiance to the fantasy genre, Pratchett made heroes of unattractive old ladies, in contravention of the convention that deems they must be young, male, and handsome. In the case of the elves, Queen Nightshade is constantly seeking to escape Fairyland and enter the Disc, whilst the King has withdrawn from active participation, content to sit in his underground sweat lodge. Although she is evil and despotic, the Queen is an active and vital force, whereas her erstwhile partner appears to have given up, content to wait until the end of time to reassert his kingdom on the Disc. The entrance to their domains are blocked by ancient stones impregnated with meteorite iron, but the Queen continues to test the boundaries, waiting for a chance to slip through and wreak havoc on the land beyond.

In his depiction of elves, Pratchett also addresses another favourite theme of class and social and economic inequity. Pratchett returns repeatedly to this theme as Sam Vimes, Duke of Ankh resits the temptations of rank and clings doggedly to his proletarian roots. He deplores the very notion of aristocracy and royalty "What set Vimes's teeth on edge was the idea that kings were a different kind of human being. A higher lifeform. Somehow magical" (Pratchett 1996: 96). Most aristocrats, in Discworld regardless of species are depicted as unpleasant and stupid, with few honourable exceptions such as

Lord Vetinari, Lady Sibyl Rankin, Dwarf Queen Rhyss Rysson, and Diamond King of the Trolls. When the throne of Lancre becomes vacant, Pratchett places a literal fool on the throne in the form of the King's former jester, Verence, who turns out to be a solemn, thoughtful, and forward-thinking monarch and an improvement on the former incumbents. It is significant then, that given Pratchett's dim view of the titled and entitled, that aristocratic euphemisms are used instead of the word elf. Elves are given the aristocratic names which are always capitalised, such as "Lords" and "Ladies", The Gentry, or names indicating their superficial attractiveness such as "The Shining Ones", "The Fair Folk", or "The Star People". These alternates are necessary because as Nanny Ogg explains, "they come when they are called" (Pratchett 1992: 63), and given their cruel nature, encounters with elves are best avoided. Although the human Disc communities have forgotten the evil oppression of the ancient rule of the elves, protective measures linger in half-remembered traditions such as spitting and touching iron after referring to them, although never by name.

As always, Pratchett's sympathies are with the downtrodden and the marginalised but his anger at the injustices of the world are always tempered with the humour and humanity that are his trademark. As his friend Neil Gaiman commented in 2014, "There is a fury to Terry Pratchett's writing: it's the fury that was … Terry's underlying sense of what is fair and what is not". The Elves are painted in such a dark light because they trample over Pratchett's most strongly stated moral law, that treating others as objects is the root of all evil. Esme Weatherwax states this clearly in *Carpe Jugulum* in her argument about ethics with Reverend Mightily Oats: "And sin, young man, is when you treat people as things. Including yourself. That's what sin is". Mendelsohn writes that the ethics of Discworld are rooted in individual autonomy (2001: 161). This view is endorsed by Noon (2010: 33), who writes "[in] Pratchett's fantasy world, real power is entirely a function of human belief and human choice". Shanahan points out that there is no one evil entity in Discworld such as Voldemort or Sauron, instead Discworld villains "are the baser elements within us all, and a wider culture encouraging uninformed prejudice and deliberate ignorance" (2018: 32). Thus, the elves are evil because they cannot understand the right to autonomy of other beings and treat others as objects and playthings for their own amusement.

Ever the humanist though, in his final Discworld novel, Pratchett gives the Elf Queen Nightshade a chance to become more humane and sympathetic. If the breaking of Prospero's staff in *The Tempest* can be seen as Shakespeare's farewell to theatre goers, *The Shepherd's Crown* (2015), may be read as Pratchett's goodbye to his readers. Certain parallels can be drawn between *The Tempest* and Pratchett's last work, including the common themes of the corrupting influence of power and the magic of forgiveness. How far this is a conscious decision by the dying author is a question that may never be answered, although it is true that previous witch novels resonated strongly with other Shakespearean plays, namely *Macbeth* (*Weird Sisters*) and *Hamlet* (*Lords and Ladies*). In *The Shepherd's Crown* Pratchett gives us the death of uber-witch Granny Weatherwax, and the taking up of the mantle by Tiffany Aching, representative of the new generation who must find their own wisdom in kindness, duty, and hard work. Thus, Tiffany teaches the Elf Queen to engage in manual work as this is the path to self-knowledge and real power – the power of compassion and self-sacrifice. As Prospero learned to live without magic, so does Nightshade, and consequently learns a different way of leadership. Eventually she is redeemed when she dies attempting to persuade the elves to live with the Disc inhabitants rather than exploiting them. In a highly symbolic act she is buried in the beloved Chalk by Tiffany and the Nac Mac Feegle, her body to eventually form part of the Disc which she was desperate to rule. This growth and change towards self-awareness and a recognition of others came about under Tiffany's tutelage. Only after she has suffered and learned empathy through performing difficult human tasks and committed the ultimate sacrifice of giving up her life can Nightshade earn redemption. This is an optimistic message; if even Nightshade can learn compassion, and the Nac Mac Feegles can learn forgiveness and restraint, then what other wonders are possible? Sir Terry has left us, but his wisdom continues to inspire and comfort.

Fernando Gabriel Pagnoni Berns

Don't Be Afraid of the Dark (John Newland, 1973)

Introduction

Supernatural creatures such as goblins, brownies, elves, and fairies were, mostly, unused monsters in horror cinema. Arguably, these little creatures are closer to being children's stuff than the stuff of nightmares. In old times, however, especially through Celtic folklore and the impulse that the Romantics gave to supernaturalism, the household spirits were very fearful menaces in Scotland, England, or Ireland, as many ballads and folk-tales recollected during the romantic period can testify.

Another probable reason preventing these little critters of being presented in film fiction more often may lie in the fact that they lack a clear definition of their forms, extent of their powers, geographical location, and general intentions and attitudes towards humans. However, some specificity can be rescued from the abyss of time. While tiny creatures such as fairies were located, traditionally, in relation with the world of nature, the space for creatures such as brownies, boggarts, or imps was, mostly, within the domestic sphere. Their domesticity and good behaviour, however, were fragile issues, as the short-tempered critters could easily turn to evil practices. While brownies "perform chores and help with any unfinished business at night" (Weinstock 2016: 61), imps were demons or spirits of mischief. Boggarts, in turn, caused mischief and things to disappear within the household. Imps and boggarts were known as pranksters (*Legends and Traditionary Stories* 1843: 83), performing little tricks within the boundaries of the domestic space. Thus, housewives had to battle against little demons when performing their daily domestic chores. The intervention of domestic demons could be a perfect excuse for less-than-perfect housewives. The lack of cooking skills and, as consequence, tasteless food or an unkempt house can be blamed on the presence of little supernatural creatures running around, meddling with household items.

It is within this scenario of gender battles that a little telefilm appears on TV: *Don't Be Afraid of the Dark* (John Newland, 1973). The film revolves around a housewife, Sally Farnham (Kim Darby), who has to battle a chauvinist husband and a parade of men trying to imprison her within the domestic space. To complicate things further, she is assaulted by increasingly evil tiny creatures (probably boggarts). The domestic space turns into a nightmare when the little creatures mess with Sally's domestic life. Soon enough, she becomes the embodiment of a "bad" wife to both her husband and society. *Don't Be Afraid of the Dark* plays with tiny pranksters with mischief in their mind as a form to engage with the politics of feminism and backlash of the 1970s.

In this chapter, I propose to make a brief outline of the status of household spirits in traditional folklore. Later, I will provide a close reading of *Don't Be Afraid of the Dark,* one of the most often-remembered telefilms of the 1970 and probably the only film uniting gender anxiety and evil fairies.

Fairies, Sprites, Household Spirits: The Domestic Sphere of Mischief

I must first address one problem: a univocal denomination for all the supernatural creatures named in this chapter is lacking. Indeed, there seems to be a lot of confusion regarding the potential differences between pixies, faeries, elves, and other legendary folk like the sídhe, brownies, daemons, elves, fairies, gnomes, goblins, hobgoblins, selkies, dwarfs, sprites, trolls, and leprechauns. As Richard Firth Green argues, there is little "to say on the vexed question of fairy taxonomy. Are fairies different from elves? or goblins? or dwarves? or pucks? or brownies?" (2016: 2). Encyclopedias on folklore address the fact that some entries overlap as, for example, barguest with bogies and brownies (Monaghan 2004: 36).

Rather than trying to make an attempt on taxonomy, I will take as umbrella terms three different denominations, all three useful for different reasons: fairies, sprites, and, more important, household spirits. The three terms progress from general to the particular. Cooke Stafford, writing in 1848 for *Hood's Magazine and Comic Miscellany*, explains that the amalgamation of global traditions, mythology, and superstitions have occasioned all diminutive beings which fable describes as related to nature "to be classed under the name of fairies" (1848: 28). The term "sprite", in turn, is an appropriate nomenclature for the *tiny* fairies. Sprites can be understood as elementals of nature.

Here, we can follow Claude Lecouteux's term "household spirits" to denominate all sprites that have a liking for living under the same roof and being attached to a family. Following Gustav Ränk, Lecouteux defines household spirits as minor deities on which the family's well-being depends. They appear in various forms but sought the society of man, and attached themselves to houses and families.

In Scottish folklore, brownies and boggarts (brownies turned evil) were such household creatures. Brownies are male beings which generally live in houses and are friendly towards humans and hard working on the domestic front, until crossed, which is when they become particularly malicious. Small goblin-like creatures wearing ragged brown clothes (hence the name), brownies do well for humans, if they are thanked, or given gifts. However, they will be very offended and disappear forever if they change their mood. Interestingly, they can be misogynists (Campbell 1999; Parsons 1964).

The issue of misogyny is a recurrent trope in some of these household spirits. Women were, historically and culturally, attached to the boundaries of the domestic sphere, while the public space was mostly the place for men. I am not stating here that the household spirits never did despicable things to men. But as household inhabitants, the small creatures share more time engaging with the women of the house than with men, the latter spending more time outside the home.

In many occasions, maidens were the butt of the supernatural jokes practiced by the household spirits. Documents from the beginning of the 17th century show domestic spirits being expelled from their places of residence. In 1615, in the house of a Lord of the Dauphiné near Valence, a brownie was playing tricks on the people living there. "One day a flask of rose water was broken by a stone, and, although it was at the feet of the Maiden of the house, the water that was in it nonetheless was all carried aloft and spilt upon her head" (Lecouteux 2000: 150). Another of the brownie's tricks was to throw objects with great force, enough to leave a person unconscious (Lecouteux 2000: 151), an act which points to an increase on violence. Playful elves in Pembroke "plagued people, by throwing dirt at them" ("Frair Rush" 1836: 520) while goblins "fright the maidens" (Thorne 1847: 982).

The household spirits' interest in keeping the house clean and neat or unkempt and dirty associated them with the role of housewives. In this reading, they can be both, the best companion or the cruelest of foes to married women.

Female Oppression through Supernaturalism: *Don't Be Afraid of the Dark*

Don't Be Afraid of the Dark predicates upon the fears and anxieties born from women's liberation movements, sexism, patriarchy, and the backlash that feminism suffered at the 1970s. In this regard, the film embodies female oppression sustained by patriarchy allegorised through the figure of little creatures tearing apart the life of a modern woman struggling to find her own voice amidst the countercultural movement denominated "second wave feminism".

Second-wave feminism refers to a collection of interconnected feminist movements that emerged in the 1960s and reached their peak in the 1970s. The second wave was primarily concerned with eliminating gender inequality and denouncing the oppression of women as structural rather than circumstantial. It was focused on a range of issues including equality between men and women, reproductive rights, and the right for women to define their own sexuality and identity.

By the 1970s, radical feminists (among other second wavers) were subjected to a daily backlash at the hands of general media, which distrusted the subversive radicalisation of the movement. Feminists were depicted in popular media as lonely and depressed women due to the shortage of men in their lives

(Faludi 2006: 1). As a result, the social advances of women were put under a media microscope and various popular culture texts, registered the political impact of the women's movement.

Don't Be Afraid of the Dark is such a text, mapping, through the inclusion of household spirits, the struggles of the main "maiden" to fit into conformity while feeling at odds with her predetermined role. Sally Farnham, the upper-class wife occupying the lead role of the film, seemingly has everything that a Western, white, capitalist woman could desire: Alex, her upwardly mobile husband (Jim Hutton), who has a promising future ahead of him, bourgeois friends, youth, beauty, and a two-storey dream Gothic house recently inherited from her grandparents. From the beginning, the traditionalism of gender binary will frame the couple and their relationship regarding the new home. While Alex mostly limits himself to live in the house, it is Sally who is in charge of decorating and remodeling the home. The film begins with Alex degrading his wife in two ways: first, he is reluctant to accept the fact that the house is beautiful. The house has come from *her* (from *her* grandparents), so to preserve his position as the home's breadwinner, he explicitly states that he finds the house only so-so. Second, as Sally ponders about their luck in getting such a beautiful house, Alex recurrently highlights the fact that the house will need a lot of repairs and Sally a lot of help. Further, while Sally is interested in redoing the kitchen (the *domestic* space), Alex is interested in the library (the *intellectual* space), thus foregrounding in concrete ways the traditional gender binary.

While Alex is busy getting promotions at work and climbing the social ladder to the upper classes (a situation that, according to his wife, is "consuming" the marriage and tearing them apart), Sally spends her day with her interior decorator Francisco Perez (Pedro Armendáriz Jr), the maid (Lesley Woods), and Harris, the handyman (William Demarest), all of them setting up the house to make it modern and classy. When making the repairs throughout the house, Sally discovers a closed room in which she finds a bricked-up chimney. She decides to open up the chimney, get it to work and turn that particular space into her office, a space – the only one within the home – that is just her own.

The handyman, however, warns her that "some things are better left alone". Even when Sally insinuates that the handyman may have fears of something supernatural – a way of diminishing him before her husband's eyes – Alex laconically mentions that "he may be right" before retiring, uninterested in the whole business. It is clear that Alex will not give support to his wife. The two men of the house reunite and, forming a close patriarchal brotherhood, deny Sally any space that can give her a sense of fulfilment. The men in the house (with the exception of the effeminate decorator), faithful to the era's warnings about the poisonous effects of freedom in women, hence, backlash any attempt of female independence.

Sally is adamant in her position and opens the chimney's ash chute with her bare hands, an action motivated by the refusal of the men to come along with her. Inadvertently, she has opened a doorway to a subterranean, chthonic landscape where creatures of old survive in modern times. Soon enough, the household spirits invade the house but only taunt and take down Sally, not Alex. All the people around Sally begin to question her mental stability, and her perfect bourgeois world falls apart.

It is interesting that the true nature of the household spirits is never fully explained. They are little bald humanoid beings, wearing ragged clothes. They are unmistakably male due to their lack of breasts, the most common signpost of femininity in any sketch indicating womanhood. These household spirits come not only from the depths of the house but also from the depths of time. Like the household spirits of folklore, their mission is to drive the maiden mad. The creatures' ragged clothes coded them as boggarts, "darker" versions of the brownies (Eason 2013: 76). The boggart "was a trickster version of the brownie, who caused destruction, tossing things about the house at whim" (Monaghan 2004: 36). While brownies specialised "in doing barn work at night: threshing, tidying, currying horses, and the like" (Monaghan 2004: 36) the boggart is the exact opposite, their main task being to put the house in disarray.

The first sign of the presence of household spirits within Sally's house will take place in the kitchen, a domestic space and the room *par excellence* of these creatures in folklore. The first concrete manifestation – the shattering of a glass ashtray – will take place in the bedroom, the place embodying an uneasy marital status. When a scared Sally mentions the ashtray, Alex is quick to interpret that she dropped it somehow, only the first of many in which he will equate the spirits' misbehaviour with Sally's incompetence.

The attack upon Sally is made through her role as the perfect housewife, undermining her position as such. In one key scene, Sally faces the little invaders during an important dinner, ruining both her reputation as the perfect bourgeois hostess and, potentially, endangering her husband's possibilities of promotion at work. The dinner was just an excuse to "impress" those in charge of giving Alex a promotion and, as he says, he wants Sally to cook something "not too simple". The house is only half finished, so Sally categorically states that "the house isn't perfect" but "I'll be the perfect hostess".

During dinner, Sally spots the little creatures among the flowers in a vase. Her expression of fear attracts the disapproving gaze of the bartender (Ted Swanson), another manifestation of patriarchy punishing women. The tension amounts to unbearable levels when, with all the guests sitting at the table, the household spirits cruelly play with Sally's napkin. She does her best to ignore them, knowing that her husband's promotion depends on the dinner and her abilities as hostess. But when Sally – and audiences alike – gets the first look at the creatures' ugly faces she can no longer suppress a scream, ruining her role as the perfect hostess and definitively disrupting the domestic space.

The presence of the boggarts may be read as signposts of the inner turmoil flooding Sally's subconscious, as she is unable to find her own voice within the home. Sally seems happy to comply with what is expected from her as a woman: a passive role within an upper-class marriage. It is obvious, however, that the Farnham marriage is not as perfect as it seems. Rather than completely fulfilling her, marriage is not the imagined perfect setup. Alex is never around and there is little communication between the couple. However, when he is at home, his role is that of undermining every single one of Sally's ideas.

The little creatures coming from the bowels of the house function as a metaphor for Sally's subconscious, a manifestation of her own self-hatred (and desire to sabotage Alex's businesses) and her attempts to "power up" in the feminist 1970s, a decade in turn filled with backlashing and fear about female empowerment. Even more, the coming of the beasts can be read as a resurgence of a common enemy of women and their attempt to improve their conditions, as the little household spirits return to put modern women in their "right" place again, as contained solely within the domestic sphere.

The use of the chimney as the doorway to another world is interesting. This particular choice (which the director could easily have replaced with any other space within the home) refers directly to the entrance of little household spirits within the home. Among the Balkan Slavs, demons try to enter houses through their chimneys at certain times of the year with evocative names: "the evil days", "the nights of the enemy", "the unblessed days", and "the days of the demons" (Lecouteux 2000: 65). Lecouteux explains that in Germany, "the souls of the dead intentionally remain close to the furnace, the habitual residence for spirits", while in Switzerland, traditions collected at the beginning of the 20th century "tell us that souls in torment stay, in front of, behind, or inside the stove" (2000: 115). The use of the chimney as the main entrance of the household spirits to Sally's home answers to both, an attempt of allegorising the female subconscious and a return to the tropes of the old folklore.

The film ends with Sally finally overtaken by the household spirits, who take her down the chimney. Sally was unable to find a balance in her life: torn apart by her desires to fulfil patriarchal mandates while looking for a more active role in life, the irreconcilable contradiction ends with her extermination, as she is forever trapped within the house, together with the little creatures that had come to help her husband – and patriarchy – to imprison her, her female claims for independence shut down together with the chimney.

Conclusion

Sally is a clear representative of the struggles of feminism during the 1970s. She tries to fulfil the role of a good bourgeois housewife, but, deep down, she is aching for something more, that exciting life promised by feminism, one built at the margins of the traditional roles for women.

In the end, Sally's voice joins the chorus of the female voices lost and unheard; her spirit forever trapped in a house that represents a life that never had been her own, but rather, the social construction of her husband. When he can no longer keep her within her assigned boundaries, the little creatures of the old folklore appear to punish her, now in a supernatural way. Both the husband and the imps want to constrain her and keep her trapped within the walls of the domestic sphere, a regression to women's "natural" status in the era in which the oral narrations about brownies, fairies, and other household spirits were told.

Don't Be Afraid of the Dark does not condemn feminism, but the backlash suffered by the movement through the 1970s (which will continue through the 1980s). The film asks its audience to connect with Sally's angst, as she can be seen in almost every scene. Furthermore, audiences share with her the vision and concrete presence of the household spirits attacking her. Both viewers and Sally know that something is really wrong within the house, but nobody believes her. She is recurrently pathologised, her warnings only seen as signs of her foolishness or mental instability.

Kristin Aubel

Rivers of London
(Ben Aaronovitch, 2011–Present)

Introduction

In *The Fairies in Tradition and Literature* (1968), Katharine Briggs concludes that fairies "are, and always have been, the Hidden People" (Briggs 1968: 210). Almost fifty years later, Stefan Ekman claims that "each attempt at pinning down urban fantasy includes at least a few examples of the Unseen" (Ekman 2016: 463). Urban fantasy thus seems to provide fertile ground to explore contemporary approaches to fairies – and indeed: "Denizens of Faerie are apparently common in at least some areas of the genre" (Ekman 2016: 460). One example is the *Rivers of London* series (2011–present) by Ben Aaronovitch, in which fairies are reimagined as liminal urban characters.

Rivers of London is an ongoing urban fantasy multimedia series consisting so far of nine novels, five novellas, one short story collection, twelve graphic novels, and one TTRPG. Its protagonist, the magical police officer Peter Grant, is, as a mixed-race man from a lower social class (see Borowska-Szerszun 2019: 10), a "hybrid in terms of class and race" (Lethbridge 2017: 237). Hybridity, "the defining feature" of the urban fantasy genre (Deffenbacher 2016: 30) not only transforms and enables new perspectives on human characters but supernatural beings as well. In Rivers of London, folk and fairy tales are "key intertexts", but only "one of several" (Binney 2018: 15), making the series part of a portion of 21st century literature that Helen Binney terms "folklore-inflected fiction" (Binney 2018: 16). In the series, contemporary London is not only populated by fairies, but also *genii locorum* (the spirits of rivers and other important places), Quiet People who live underground, sentient foxes, ghosts, vampires, and others. Together with magical practitioners, they form the so-called demi-monde or half-world. The historical context of this term – and its pejorative connotations – are made explicit later in the series, when it is specified as "an old euphemism [...] for any sexually active woman who failed to conform to the strict patriarchal gender norms" (Aaronovitch 2021: 65). Supernatural beings are thus ascribed a half-existence outside of societal norms. This conception is, however, challenged throughout the series, especially by the "curious yet open-minded" Peter (Borowska-Szerszun 2019: 21).

The term fae is used by various characters both to encompass all beings that are "different" (Aaronovitch 2013: 152) and as a synonym for fairy (e.g. Aaronovitch 2012: 148). This naming chaos is consistent with fairy folklore (see e.g., Simpson and Roud 2000b: 115, 116; Silver 2008b: 321). The series thus continues the folkloric discourse on a meta-textual level, as "consistency or logic is not to be looked for in folk tradition; for it is not one voice that transmits it, but many" (Briggs 1968: 54). This multiple discourse is further amplified through pseudo-intertextual references: Historical authors of supernatural fiction such as John Polidori (Aaronovitch 2011b: 136) are claimed as scientific authorities on fae.

In this chapter, I will first analyse the representation of fairyland in *Rivers of London* to then compare two fairy characters: Molly, who was born in fairyland and is enslaved in the city, and Zach, an urban native with clear connections to the demi-monde who rejects any attempts at classification. Both characters thus represent different strands of contemporary urban liminality.

Dealings with Fairyland

The vast majority of media in the *Rivers of London* series takes place in the urban setting of London (see also Borowska-Szerszun 2019: 6). Notable exceptions are the fifth novel *Foxglove Summer* (2014) and the graphic novel *The Fey and the Furious* (2019–2020),[1] written by Aaronovitch and Andrew Cartmel. In both texts, a rural setting is introduced that

---

[1] This title introduces yet another spelling of fae, keeping the inconsistency consistent.

then opens the way to fairyland. In the graphic novel, this is the beach of Southend (Aaronovitch and Cartmel 2019–2020: #2, 24–6). *Foxglove Summer* takes place in a village bordering on Wales (Aaronovitch 2014: 5, 29). The connection between the fae and Wales was already hinted at in the second novel (Aaronovitch 2011b: 142) and is consistent with folkloric observations: "English fairy beliefs are less lively, less documented and less profuse than those of Scotland, Ireland and Wales" (Briggs 1957: 279; see also Briggs 1961: 301).

Foxglove Summer deals with a more or less classic changeling plot: abduction, substitution, chaos, suspicion, remedies, return (Conrad 2008: 179). There are, however, a few important subversions: The exchange of the human with the fairy child is only noticed after the children are changed back and the human child that grew up in fairyland is perceived as the "false" one: She is named "not-Nicole", turning her into a negation of her sister[2] (e.g. Aaronovitch 2014: 373). The mother insists on her familial bond with the changeling instead of treating her badly, as was often done historically with supposed changelings (Simpson and Shroud 2000a: 53): "Blood isn't everything – I want my daughter back" (Aaronovitch 2014: 357)." Furthermore, it is the human child that is described as an "evil little strop" (Aaronovitch 2014: 374), making her the "mentally abnormal, and ill-tempered" (Simpson and Roud 2000a: 53) one. In opposition to folkloric tradition, nurture is thus represented as more important than nature. In keeping with the idea that "[t]he concept of exchange lies at the heart of changeling tales" (Conrad 2008: 179), Peter tries to solve the situation by exchanging himself for the two children (Aaronovitch 2014: 360–2).

The "High Fae", as the denizens of fairyland are later named (e.g. Aaronovitch 2018: 81), are described as "[h]uman shaped but tall and thin, with long delicate faces and hands and black eyes" (Aaronovitch 2014: 357), deviating from traditional folklore as "[m]ost of the English courtly fairies are small in size" (Briggs 1957: 271; see also Hahn 2015: n.p. and Silver 2008b: 321). Their movement is otherworldly, "as if they were standing still and the world was obligingly rearranging itself around them" (Aaronovitch 2014: 360), implying a position of power over the natural world. They (Aaronovitch 2014: 372),

---

2 Nicole and not-Nicole are actually half-sisters, since they share the same human father.

as well as other magical beings (Aaronovitch 2011b: 136), are, however, as in folkloric tradition (Briggs 1957: 275), vulnerable to iron.

Fairyland (or faerie) is not located in a hidden physical space such as underground or under water (see Briggs 1957: 273; Silver 2008a: 319; Simpson and Roud 2000c: 116) but in a "parallel dimension of some kind" (Aaronovitch 2014: 366) and thus theoretically even more out of reach than traditionally (Simpson and Roud 2000c: 116). Instead, the realm is in contact with and influenced by the human world, as can be seen for example with its castles. These castles are not historical buildings but designed to appeal to modern nostalgic sensibilities, becoming "pretend castle[s] built to indulge the fantasies of a ruling elite" (Aaronovitch and Cartmel 2019–2020: #4, 13). Human imaginaries thus already have a disruptive effect on high fae identity. The spread of human civilisation is even more detrimental, violating the sanctity of fairyland: "The Roman road. Those imperial fuckers had put their mark on the landscape, all right. Even to the point where it impinged onto fairyland" (Aaronovitch 2014: 369). Various fairy courts use these connections to the human world to their economic advantage, exporting unicorn horns (Aaronovitch and Cartmel 2019–2020: #3, 20–1) or selling their own people into slavery (Aaronovitch 2018: 315). Even the fae of fairyland thus become implicated in urban modernity.

Urban Fairies: Molly and Fae Trafficking

Molly is introduced in the first novel as the housekeeper of the magic police headquarters, the Folly: "I looked over and saw a woman gliding towards us across the polished marble. She was slender and dressed like an Edwardian maid [...]. Her face didn't fit her outfit, being too long and sharp-boned with black, almond-shaped eyes" (Aaronovitch 2011a: 80–1). Further descriptions focus on her animalistic and dangerous character, for example "spider-like motion" (Aaronovitch 2011a: 373) or "as flexible as a snake" (Aaronovitch 2011a: 274), dehumanising her even more.

Molly's backstory is slowly revealed throughout the series: She was part of a group of female fae who were "traded by [their] queen for something valuable" (Aaronovitch 2018: 315). The group was quickly separated and put to work in London for several decades (Aaronovitch 2018: 315). In 1911 (Aaronovitch 2011b: 215), Molly was rescued from what the police identified as sex trafficking (216). She was perceived as a "strange young European girl" (216) and deemed "too abnormal to be fostered" (217) and was thus brought to the Folly. Since then, she has not left this building (214). One hundred years later, during the events of the series, she has the appearance of an adult. Her slow ageing fits some folkloric accounts: "As to their span of life, occasionally fairies are said to be immortal, always very long-lived" (Briggs 1957: 273). This is used to her disadvantage, though, as her longevity in combination with her superhuman strength and otherworldly appearance, once magically enslaved, turn her into a valuable sex and menial worker.

In *Foxglove Summer*, Peter realises through the similarities with the high fae in appearance and mannerisms, what Molly's origin must be: "So Molly was fae or, even better, this particular kind of fae – whatever this kind of fae was" (Aaronovitch 2014: 360). Molly is unable to speak, which the high fae in fairyland are usually not (e.g. Aaronovitch 2014: 371), indicating mental or magical trauma. Her inner life is thus mostly inferred from her behaviour and interpreted through the perspectives of other characters.

Peter's mentor and supervisor Nightingale believes that Molly does not leave the Folly because "she's frightened of what's out there" (Aaronovitch 2011b: 214) and that people like Molly "were changed by magic, or they were born into lineages that have been changed. And as far as I know, this leaves them – incomplete" (Aaronovitch 2011a: 215). Nightingale's explanations have to be taken with a grain of salt, however, since they are tainted by racism (Borowska-Szerszun 2019: 12–13) and "postimperial melancholia" (Borowska-Szerszun 2019: 20). Peter questions and disrupts this way of thinking:

> Nightingale, despite literally being a relic of a bygone age, had learned to modify his language around me because when I'd looked into the literature, the most common terms started with "un-" – unfit, unsuited, undesirable – and behind them came the terms starting with "sub-". However, with a bit of running translation, it was clear that "incomplete" people like Molly were vulnerable to abuse and exploitation by their more powerful brethren, and by practitioners with no moral scruples (Aaronovitch 2011b: 215).

The series thus combines the pseudo-historical othering of magical beings with a subversion of historical othering of human beings justified through folklore (changelings), making visible how deeply entrenched othering is in our knowledge-making ("showing us what we do not want to see", Ekman 2016: 465), but also suggesting that it can be overcome with empathy. This discourse extends beyond human beings in the genre of urban fantasy, now including the (in)humane treatment of magical beings, which even the hybrid Peter is not able to fully resolve.

Over the course of the series, more victims of fae trafficking are introduced. The most prominent ones are the "Pale Lady" (Aaronovitch 2011b: 130) and Foxglove (Aaronovitch 2018: 290). The Pale Lady, conspicuously the only one of them without an individual name,[3] remains monstrous – indicated by descriptors such as "disturbing" and "terrifying" (Aaronovitch 2011b: 277) – and is killed in a fight with Peter (289).

Molly and Foxglove, however, become less other but stay liminal. Foxglove is able to tell her life story through pantomime, which still requires interpretation, but is all in all a communication success (Aaronovitch 2018: 311–26). Both fae are able to embrace new identities and purposes after their enslavement; Foxglove as an artist, and Molly as a housekeeper. They embody and communicate these identities through their clothing choices: "I realised that the loose top she [Foxglove] was wearing was a linen artist's smock – in fact, all she needed was a beret and the cliché would have been complete. I thought of Molly and her Edwardian maid's outfit and wondered if the costume was significant" (Aaronovitch 2018: 307). Their clothes thus become significant markers of their fae identity.

A return to fairyland is never discussed, making the city, the site of their trauma, their new home. Their involuntary presence in the city is already a transgression in the sense of blurring the border (Klapcsik 2012: 14) between human and fae society. Molly literally lives on the threshold, successfully inhabiting the limen (Klapcsik 2012: 7, 121), as she never leaves the Folly. She can connect with it and the whole city of London on the level of time, though: She is able to access the "institutional memory" of the place of the

---

3   Foxglove's name as well as the title *Foxglove Summer* are very likely references to the fact that foxgloves are "universally reputed to be fairies plants" (Briggs 1957: 85).

Folly (Aaronovitch 2011b: 107) and to act as a conduit to London's memory plane (Aaronovitch 2011a: 362). Molly thus claims her position as fae within the history of the city in spite of trauma and magical vulnerability.

Urban Fairies: Zach as Urban Drifter

In Zachary "Zach" Palmer's first appearance, he is described as weird but is, in contrast to Molly and her sisters, not immediately recognisable as other or supernatural (Aaronovitch 2012: 46). His parentage is only revealed later: "My dad was a fairy" (Aaronovitch 2012: 141). He is subsequently described as "demi-fae" or "half-fairy" (Aaronovitch 2012: 148) or "goblin" Aaronovitch 2012: 292). Goblin is a "general term for fairy creatures of malicious or evil nature" (Simpson and Shroud 2000d: 146). Zach rejects all of these labels, making clear that this classification system is imposed on magical beings by human outsiders (Aaronovitch 2012: 342). Folkloric accounts would agree that it is "unwise to name the fairies" (Briggs 1957: 276).

In consequence, Zach is "the neighbourhood odd job guy" (Aaronovitch 2012: 292) for the demi-monde. His employment, loyalties, and general relationships are constantly in flux. Although he does work for the river spirits, he describes them as "half a bunch of stuck-up cunts" (Aaronovitch 2012: 342), revealing class tensions within the demi-monde. In contrast to them – and Molly – he is not bound to a certain place: He is "less powerful" but "more independent" (Aaronovitch 2012: 148). His access to magical urban spaces is chaotic and shifting: He is able to visit a pub that is "strictly fae" (Aaronovitch 2013: 221) but is a "persona non grata" at what Nightingale calls the "goblin market"[4] (Aaronovitch 2012: 149). He works as intermediary for the Quiet People but is not allowed into their society either (Aaronovitch 2012: 350–1). His disregard for any boundaries (together with his talent for lock-picking) not only enables him to enter locked spaces (Aaronovitch 2013: 219) but

---

4   The goblin market is likely a reference to Christina Rossetti's poem (1862) of the same name.

also excludes him from the larger urban society (also see Borowska-Szerszun 2019: 20, 21).

Zach is the ultimate "go-between figure" (Lethbridge 2017: 246), even more so than Peter. His liminality is marked by "aimless wandering" (Klapcsik 2012: 166), making him an "urban vagabond" (Klapcsik 2012: 166), the hybrid who belongs everywhere and nowhere. Since he grew up in a children's home (Aaronovitch 2012: 339), Zach cannot return to a stable family home either, just as Molly or her sisters cannot return to fairyland. As urban native and drifter, the whole city is and is not his home at the same time.

Conclusion

The urban fantasy series *Rivers of London* includes fairy folklore as one of its intertexts. Fairyland is accessible through and thus associated with the countryside. Its inhabitants, the high fae, are not an isolated people, but exchange ideas, goods, and people with the human world. Consequently, the fae are included into the worldmaking and themes of urban fantasy, especially in the context of liminality and hybridity. Both the enslaved high fae, such as Molly, and urban natives, such as Zach, live liminal lives. They are classified as other and only ascribed a half-existence. Nevertheless, they manage to assert their place in the city: Molly claims her identity as the housekeeper of the Folly, and Zach is a restless drifter, always finding a place where he is able to stay for a while. In this series, the urbanisation of fairies thus does not resolve contradictory folkloric accounts but instead amplifies the characterisation of fairies as in-between figures inhabiting the limen.

Part V
Across Cultures

Clay Franklin Johnson

Poem: The Moonlight Meeting of Lord Ortho and Setareh

She awoke with the last days of spring, bringing
New night-music among the ruined woodland
As light footsteps fell softly upon dead flowers,
Awakening new life upon the midnight hours

Beneath her feet slithered beautiful things,
Creatures wrapped with luminescent serpent-scale,
Flickering like ghost-candles of black opal—
The coldest glow of the will-o'-the-wisp flame

Within the woodchips of once-living bark
Crawl tree-worms reflecting cold stars, swimming
Delirious through drippings of Venetian glass,
A liquid lust from teardrops of blue goldstone

Within the littered decay of dead flowers
Of livid purples and corrupted blues
Lives a new light, a new perfection of Night,
Living for new death in ethereal hues

Above her floating hair floats pale whisperings,
Flutterings of shadow, gossamer-winged
Sylphs, spirits of the diaphanous air,
Silken delicate they manifest fine & fair

Born from death, from the dying of the light,
From the withering of flowers & ice,
The frosts of starlight, bleeding old life
With a liquid kiss it consumes itself in new life

New life, new to this world, but not the *Other*,
Not to the mysteries of celestial light,
Nor to the songs of starlight—No, its bright life
Has been buried deep, deep within oceans of time

Yet, in recent times, such life has been nightmared,
Falsely called evil by obsequious minds,
Bedeviled with names of demons & devils,
And blamed for all the world's wrongs & ills & troubles

And even she, the one who walks light-footed
Atop scattered tree-bark and flowers murdered
By the hypocrisy of artless men,
Has suffered the misogyny of unjust names:

Mother by first pagans, and Goddess Divine,
Then Astarte, Pandora, Aphrodite,
Child-eating Lamia, Eve the snake-friend,
And La Belle Dame sans Merci with her *wild, wild eyes*

She was born from a whisper, a tree-shadow,
A lingering of dark from the Evening Star—
Yet she absorbed the starlight, consumed its dreams,
And became The Fated Star, or green-eyed Setareh

And thus, it is she who walks in the starlight,
Not in beauty, but in cold philosophy,
In the human ruin of ancient plant-life
In pursuit of sick lust for life of luxury

Long did she linger in the cold starlight,
Misty-eyed with deep despair, crystalizing
Each emerald of green within her spectral eyes
With a mystery of unfathomable life

Until she peeled off her skin of star-shadow
And became the moonlight, drowning out all stars
With a brilliance of pure *silver stellaria*—
And cold & sick became green-eyed Setareh

Poem: The Moonlight Meeting of Lord Ortho and Setareh

Silently sorrowful she stood as if captured
In a single beam of haunted moonlight,
Changed, like a face behind casketed glass,
Entranced with her own self-absorption for wrath

Though cancered by moon-silver, she remained
Ever beautiful, as if the pale new light
Revealed a new magic within her eyes,
Unveiled only in moonlight like enchanted gems.

Shane Broderick

"The Otherworld" (Anon., c. 1200 BCE–Present)

Fairy Revenge in Irish Folklore

Ireland is truly blessed with having a wealth of folk traditions stretching back millenia, many of which are still observable in the everyday life of its inhabitants. It has one of the largest, and best, folklore archives in the whole of Europe (Gilligan 2023: 26). This outstanding collection (digitised on Duchas.ie) is the fruits of the Irish Folklore Commission (IFC), who in 1935 started a concerted effort to collect and catalogue the vast corpus of rich folklore that abounded the entire country. The IFC was fearful that in a quickly modernising world there was a very real chance that these folk traditions might die out, were they not collected. They began with what is now referred to as The National Folklore Schools Collection (hereafter NFSC).

This project was used as a test bed to find the best areas with the richest folklore and the method of collection was ingenious. Between 1937 and 1939, over fifty-thousand school children from across the Republic took part in the project, and instead of homework, they were tasked with interviewing older people in their communities using pre-selected questions with the aim of recording all manner of folklore and traditions. This later resulted in full-time and part-time collectors travelling around the country, visiting communities, and seeking out storytellers[1] and other living repositories of tradition who

---

1   Storytellers usually came in two forms; a *Seanchaí* who handled legends and shorter tales, and a *Scéalaí* who specialised in longer, multi-episodic tales that might take multiple nights to recount in full. The latter usually had hundreds of these tales in their repertoire. An astounding feat considering the fact they were mostly illiterate.

lived there, to rescue the material that was in danger of disappearing (Gilligan 2023: 26). It might come as no surprise to anyone familiar with Irish folklore that there is a abundance of material relating to fairy lore, given how pervasive the belief in these creatures is, even up to modern day. This chapter will look at how dangerous it is to cross these denizens of the Otherworld.

Unfortunately, due to modern pop culture and Victorian romanticisation, the presiding image that comes to mind as soon as the word "fairy" is uttered is the twee, benevolent, winged, tinkerbelle-esque sprite. The Irish fairy is completely different. Generally, you would not be able to pick them apart from everyday humans. They average five feet or higher (but can seemingly change their height at will) and have a range of hair colours (Daimler 2022: 8). Their more human appearance can often be a result of a magical glamour, and when this is broken, they might appear as more wizened and ugly (Green and Lenihan 2004: 50). They might also look "of a certain time" in terms of their outdated clothing.

And then, the question of what they are arises. There are two main origin stories that are the most pervasive. The first is that they are the ancient gods of Ireland, the *Tuatha Dé Dannan* (Gilligan 2023: 16), diminished in power and influence having been driven underground and into the hills following their defeat by the Mileaseans. The second is that they are half-fallen angels (O Brien 2021: 27), who could not choose a side between God and the Devil. As a result of this, they were cast out of heaven, but not all the way to hell, so they ended up scattered across the earth. The sheer prevalence of them in Ireland was due to the belief that more of them landed in Ireland than anywhere else. So many that they were "as thick as the hairs on a dog's back" (Green and Lenihan 2003: 38).

The renowned and respected *Seanchaí*,[2] Eddie Lenihan (Green and Lenihan 2003: 25–33), collected a fantastic account in relation to the fallen angel version, which explains why the fairies are antagonistic to humans. A fairy tasks a priest with asking a dying man what will happen to the fairies on judgment day. When they are told that they will not see salvation, they are enraged and proclaim that "[f]or five thousand years we have walked the roads of Ireland, playing music and not harming anyone, but no more, Anyone

---

2   Traditional storyteller.

caught out after dark will be attacked". A caveat should be added here that they are not entirely malevolent or antagonistic, but it has to be remembered that they are not human, have their own agency, and are morally different to us.

They come from the Otherworld, known in the Irish language as *An saol Eile* (Daimler 2022: 3) or *Alltar*[3] (Ó Dónaill 1977). This, as the name would suggest, is a complete other world that exists alongside our own, full of varied supernatural (and possibly immortal) beings, and that world itself displays supernatural characteristics (Carey 1987: 1).[4] This Otherworld[5] exists alongside our own, albeit magically hidden from our eyes and can be accessed by going underground, across or under water (Carey 1989: 31) or by entering through a burial mound known as a *Sídhe* (pron: Shee). Entry through the latter was more likely on supernaturally active times of year, such as the festival of *Samhain*, where the magical barrier hiding them is rendered useless and all the *Sídhe* mounds are open (Carey and Koch 2003: 199). The Otherworld is made up of many different regions (O Brien 2021: 53), with the "fairy" otherworld being a part of that.

The term *Sídhe*[6] as mentioned above can mean the otherworld mounds, but it can also mean the otherworld itself, and the people who inhabit it. It is one of the main circumlocutions and kennings used when referring to the Irish fairy. The word "fairy" was typically seen as a taboo for fear of offending or invoking them, so creative workarounds were used to avoid retribution. The most common in the Irish language were simply *Aos Sídhe* or *Daoine Sídhe* (Daimler 2022: 2) both essentially meaning "People of the fairy hills" or "People of the otherworld", but there are many more in both the Irish and English language. Some of the names lean more towards honorifics intended to placate them, as opposed to being indicative of their nature, such as "*Na Daoine Maithe*/The Good People", "*Na Daoine Uaisle*/The Noble People", "The Good Neighbours", "The Other Crowd" etc. (O Brien 2022: 19). Even though their world is largely hidden from us, their abodes can be found dotting

---

3   The other life, the next world.
4   Such as time bending, unageing, and abundance.
5   I cover how the belief in this otherworld changed throughout the centuries in my blog post (Broderick 2019), titled "Evolution of the Irish otherworld" at: <https://irishfolklore.wordpress.com/2019/07/26/the-evolution-of-the-irish-otherworld/>.
6   This is typically rendered in modern Irish as "Sí".

the landscape and any transgression towards them usually results in severe retribution.

These dwellings appear to our eyes as physical elements on the landscape in two main forms:

1. "fairy forts" (referred to in Irish as *Rath* or *Lios*). These are the remains of medieval enclosured dwellings[7] that dot the landscape and over the centuries have built up the reputation of being the domain of the fairies. To cut down or interfere with these in any way would lead to misfortune or death to the transgressor or their family (Gilligan 2023: 18).
2. "Fairy trees", referred to in Irish as *Sceach* or English as a lone bush or lonely bush. These trees often stand out as being particularly gnarled or unusual looking and might fruit at weird times of the year. Similar to the forts, these will be given a wide berth and are often left alone, and undisturbed in a field, regardless of how much of a nuisance they are.

In terms of forts, the belief is so strong that in 1911 when a road was meant to be built through one, the "people almost rose up in rebellion" to change the course of the road (Gilligan 2023:18). Some of these forts are so synonymous with fairies that they have left their mark on the place names.[8]

The forts were almost always left undisturbed and were never ploughed by the farmers whose land they inhabit (NFSC Vol. 0028: 0232). Lenihan records a fascinating tale where a man fires a shotgun into a fort thinking it is a goose he is firing at (2003: 138–43). Of course, it is instead one of the fairy folk in the form of a bird he has shot. The person he was with admonishes him, but he pays no heed. He is found the next day with all the bones in his body twisted. He is visited that night by the fairies, who are able to pass through the walls carrying a coffin. Inside lies the body of the fairy that was shot. The man is tasked with removing all the buckshot before sunrise, or they will kill

---

7   It is estimated that 45,000 examples of these survive.
8   Lisfarbegnagommaun ('*lios Fear Beag na gCamáin*' The Fort of the Little Men with Hurleys) and Lisnagunnel ("*Lios na gCoinnle*" The fort of the candles). The first is a reference to a fort where the fairies played the national sport of hurling, and the second one where lights were seen, as if invisible people were carrying candles (Broderick 2023: 89).

him. Another man who tried to dig into a fort was blinded in one eye by a bramble bush (NFSC Vol. 0956: 207) and the same fort was known for having any animal that went inside it disappear. The loss of one eye could be seen as merely a warning, but another man (Green and Lenihan 2003: 124) ended up dead a week after when his shed fell on him.[9]

Similar to the forts, any form of interference with the "fairy trees" can result in serious harm. Sometimes they will be gracious enough to give you a warning. The axe you are using might fly out of your hand, or you might get knocked to the ground (NFSC Vol. 0028: 0232). Other times, it might be a bit more serious, like the man who had his hand wither and his subsequent death a few days later (NFSC Vol. 0834: 326). Some people have been known to spontaneously start bleeding from the nose after cutting into the branches, presumably as a warning to rethink what they were doing (NFSC Vol. 0145: 149). Unnaturally big thorns might end up beneath the skin as a result of cutting the tree, and the only way to get rid of them is by returning the tree to its original spot (NFSC Vol. 0168: 284). Fairy activity in or around these trees was fairly common, with fairies dancing or playing music by them, especially at supernaturally active times of the year like midsummer (NFSC Vol. 0510: 021; Vol. 0546: 092). One fantastic account from the school's collection (Vol. 0574: 050–1) tells how a group of field workers were picking potatoes in a field, in close proximity to a lone bush, when all of a sudden, a group of hooded women in cloaks entered the field. The men immediately recognised them as fairy folk, so they kept their distance lest they fall afoul of them. The women appeared to be taking all the potatoes and carrying them in their aprons, and when it seemed like there were no more to be picked, they left the field. The men were shocked to find that not a single potato had been taken.

Another man fell afoul of the "other crowd" when he dug up a "lone bush" to build a barn. For three nights in a row, he heard crying coming from the tree. His curiosity got the better of him and when he went to investigate, he found a few women dressed in white crying by the tree. They addressed him by name and told him he was going to suffer for what he had done. The following morning, he awoke paralysed and unable to move. A local witch

---

9   The length of time is irrelevant in Irish tradition. Someone could die 30 years later, and it would still be blamed on the time they transgressed against the fairies.

informed his wife that returning it to where it was would bring him back to normal. This was done, he recovered but the curse was transferred to his wife, who sickened and died (NFSC Vol. 0013: 215–70).

The fear people had of removing those trees was so strong that in 1999 when road builders attempted to remove a fairy tree (one well known as being the meeting point of the Munster and Leinster fairies for their faction fights[10]) to build a bypass, public outcry caused them to alter the route of the road around it (Gilligan 2023: 19).[11]

Fairy dwellings tend to all be connected through invisible paths that cross the landscape. It was also believed that they used these paths to move between their houses, especially at certain times of the year, like *Bealtaine* and *Samhain* (Daimler 2022: 20).[12] Building on these paths could be detrimental to the well-being of everyone in the house and in some cases lead to the death of livestock, abandonment of houses, or the destruction of the buildings themselves,[13] which can manifest as night disturbances in the house[14] resulting in particular doors or windows having to be always left open. All of these could be avoided by checking if the house is "in the way" before building, which could be done by stacking sods of earth, stacks of stones or putting willow rods in the ground, tracing out the outline of the house (O Brien 2022: 91–4). If the house was "in the way", these would be moved overnight. The process was repeated until they were still in place, thus avoiding revenge from the "other crowd". One story collected by Lenihan (2003: 146) tells of a dog who was always seen "shadow boxing" things that were not there and it kept escalating until the dog was found dead, while another tells of a shed that was magically moved with no damage on a calm night by a fairy wind (Lenihan 2003: 165).

10   For information of human faction fights see the blog post (Broderick 2019) titled "Irish Stick Fighting and Faction Fights".
11   This bush is known as the latoon bush. The *Seanchaí* Eddie Lenihan raised international interest in saving the bush from destruction.
12   Bealtaine (May Day) and Samhain (Halloween) are two of the four cross quarter days in the Gaelic calendar. They are supernaturally active times of year where extra protection is needed to stay safe from the fairies.
13   The walls may be knocked down or the roof might be torn off.
14   Such as banging noises, or the sounds of plates being broken.

Animals or humans trespassing near their abodes or crossing these paths can be hit with "elfshot"[15] or may suffer "elfstroke" (*Poc Sídhe*), Both of which will result in sudden illness, change in behaviour or seizures in either animals or humans (Daimler 2022: 23). It could also result in people gaining otherworldly, esoteric knowledge or second sight at the cost of mental or physical well-being (Daimler 2022: 25). The usual methods, such as using holy items, might be enough to reverse the damage, but in the case of "elfshot", small arrowheads called "elfstones" might need to be used (NFSC Vol. 0086: 9; Vol. 0155: 05–41). People might also find themselves falling afoul of the *Fóidín Mearbhaill* (stray sod) or the *Féar Gorta* (Hungry grass). The former can end with you "going astray", like one lady who was on her way to a wake (NFSC Vol. 0101: 125), fell asleep while walking. She managed not only to keep walking all night but to walk across the water without getting wet. It was only the rising sun that woke her. Unfortunately, she started shivering, took to bed and never recovered. The usual method of escape was to turn your clothes inside out, thus breaking the spell (although you have to be awake unlike the unfortunate lady mentioned above). The latter mentioned above, the "hungry grass can" also be deadly. Usually, when someone stands on one, they are overcome with a great weakness in the limbs, and a terrible longing for food (regardless of how long it has been since they ate) which might result in the person falling over and wasting away if nobody is there to help: This can be counteracted by always carrying a piece of bread in your pocket (NFSC Vol. 0130: 521).

The warning in the stories is that, especially when in Ireland, do not fall prey to the modern notion of the Irish fairy folk being tinkerbelle-esque, twee little creatures. There is a reason these tales have been preserved and continue to be used to teach children how to stay safe when dealing with these denizens of the otherworld. It is the same reason why a plethora of customs and traditions have developed, stretching the whole way back to the earliest written myths of the country to keep ourselves, our families, and our livestock safe. Heed the centuries of warnings and protect yourself, lest you fall afoul of the Sídhe!

---

15   These elfstones or elfshot are usually found in the form of neolithic arrowheads and were implemented in a number of different cures, especially in curing people and animals who were "fairy struck".

Muskan Dhandhi and Suman Sigroha

Raja Ka Sapna [A King's Dream] (Anon., n.d.)

Sisterhood and/or Lesbianism in Haryanvi Folktales

A northern India state, Haryana possesses a rich social-cultural fabric where much of its folklore is yet to be discovered, documented, and analysed. Haryanvi folklore denotes the socio-cultural reality of its society and reflects various problematic societal notions the state encounters on a day-to-day basis, such as honour killing, dowry, female infanticide, etc., thus subjecting women to extreme marginalisation and consigning them to peripheral spaces.

The oral narrative of a society often depicts its reality in some form or another. When narrated and orally transmitted across Haryana (a northern state in India), folktales often display fairies amidst violent spaces as they are visualised within patriarchal, misogynistic, and hierarchical narratives, locating them as submissive and marginalised characters. Since fairies act as a creative and folkloric representation of Haryanvi women, it is no wonder they are often translated across violent spaces within oral narratives. Here, their bodies become sites of fragility and rupture, as patriarchal and misogynistic forces portray these fairies (which, this chapter posits, are a representation of Haryanvi women) as subjected to violence, much like how Haryanvi women suffer. Any resistance on their part is met with more cruelty or dismissal.

While fairies (and women) bear bruises inflicted by forced domestication and erasure of their identities, some folktales also create textual spaces where such domestication and erasure are challenged, resulting in the creation of new identities. Few oral narratives envision fairies (or women) as a collective, with a collective identity, to resist such hegemonic violent forces. One is led to wonder about the scope of any same-sex relationships, and if so, its nature – is it sisterhood or lesbianism – that might be absent in this narrative. This chapter proposes to study such relationships between fairies in Haryanvi folktales. It plans to investigate and analyse whether these two notions can co-exist or is it essential for only one of them to exist, and do so by rendering fairies (and women) and their bodies as sites illustrating violence, fragility, and rupture, alongside examining the role of such oral narratives in reflecting the realistic notions the Haryanvi society possess.

Women in Haryana

Primarily an agricultural state, Haryana is infamous for its pronounced son preference that is aided by modern technological advancements like ultrasound machines, which are supposed to help detect medical issues but have instead been used to detect the sex of the foetus. Sex-selective abortions or female foeticides, female infanticides, and abandonment of or inattention to the female child (Ahlawat 2013; Sudha and Rajan 1999) have resulted in a grossly skewed sex ratio of 830 girls per 1,000 boys as per the Census 2011 data. This son preference is a result of some socio-economic factors that include, among others, a dependency on land for socio-cultural status – the larger the landholding, the higher the social standing therefore having daughters may lead to more claimants to the land, leading to smaller landholdings for the sons; dowry – daughters need increasing amounts of money to marry off while daughters-in-law bring dowry; cultural rituals – births, marriages, and festivals where it is socially expected to bring gifts to the sister's or daughter's family; idea of family honour – a legacy of colonial times that bestows the women of the house with sole responsibility of bearing or losing family honour, discussed below in some detail. Despite the existence of laws that give equal rights to girls in the ancestral property of their parents, professing such a claim still leads to social ostracisation of the women doing so (Ahlawat 2012). This hierarchical familial and socio-economic organisation of society, positioned firmly on gender, consigns Haryanvi women to the periphery, where they are often subjected to differing degrees of violence, injustice, and prejudice regularly. Several norms and restrictions, especially about marriage, caste, and clan, across communities in Haryana must be complied with, and any disobedience is met with actions like disowning and/or killing (especially a woman). In case a woman is allowed to live, she may be either forced into marriage with another person or disowned by both family and village for committing a breach of *izzat* (honour). While this brief section tries to show the position of women in Haryana, the following section discusses a representative folktale and discusses the notion of sisterhood within Haryanvi folklore.

Raja Ka Sapna (A King's Dream)

Raja Ka Sapna focuses on a King's fascination with his dream. The narrative progresses as follows:

A King, who had recently got married, told his wife that he often goes for a morning hunt and sleeps for an hour after returning. He requested that she give him hot water and food once he returned, and wake him up after one hour's sleep. The queen religiously followed his orders. One day, the King had a dream; he was standing in Lord Indra's courtyard and was having a good time. The fairies were about to begin their dance when the King's dream was interrupted by the Queen. The furious King disowned her and told her to leave. The castaway Queen went to a nearby village, where she met a hermit, who took her in after listening to her story. After some time, she gave birth to a boy, who was predicted to be the next King by the local priest. Years pass, and one day, he goes to a fair in a far-off town. Being jealous, other children tricked him into treading a different path. He chanced upon a well where four fairies were playing *Choupad*.[1] Upon seeing him, the four fairies went inside the well but left the *Choupad* outside. The boy picked it up, and it was subsequently gifted to the King, who gave it to his ruling Queen. However, she demanded that the *goty (dice)* be also brought. The boy went to the same well

---

1   An ancient Haryanvi game closely resembling the game of snakes and ladders.

again, this time he followed the fairies inside the well requesting the *goties*. The fairies happily gave them to him after extracting a promise from him to return. The boy left only to return as promised. The fairies transformed him into a *tabalchee* (one who plays *tabla*), who pleased Lord Indra to get a boon. In a turn of events, the fairies helped him reach the King's courtyard with his mother to get her revenge. A surprised king remembered his dream and enquired who they were. The former Queen replied, "Have you forgotten that you just saw the dream, look at my son who made you reach the heavens!" The King apologised to the Queen, the boy became the heir apparent, and the King and the Queen went on a pilgrimage.

In the fairy tale, while female characters appear as the Queen (disowned and ostracised by the King) and the fairies (residing near the well and in Lord Indra's courtyard) only, there are many male characters in the text, such as the King, the hermit, the boy, men in Lord Indra's courtyard, and Lord Indra. Men, therefore, obtain the most textual expanse within the narrative, whereas women are marginalised and appear as objects crafted for men's leisure and pleasure. The fairies within the well and the courtyard do not bear any names and surface as marginalised characters within the dominantly male spaces. In addition, they are portrayed as and remarked upon as objects of desire and as the epitome of beauty; the only performative constructs the fairies offer is where their bodies are portrayed as a site to exhibit leisure, pleasure, and performance, endeavouring to gratify the men within Lord Indra's courtyard and maybe the men present nearby the well. It is intriguing to note that initially when the boy happens to chance upon the well and sees the fairies dancing and having a good time on their own, in each other's company, the fairies go into hiding and refuse to come out. For them, it is an apparent infringement upon their private space or a blatant intervention into their privacy. If one strives to examine this relationship between those fairies, and tries to understand it, it can be read as sisterhood and perhaps even lesbianism, although the narrative stops short of exhibiting it. This sisterhood is made more apparent later when finding out about the Queen's plight, they devise the plan to help her get her revenge, despite not actually knowing her.

This sisterhood can be inferred from the fact that the fairies reside in the well together, share a close-knit relationship with each other with surprisingly no interference from the other gender. The well becomes the physical space that contains them; the underground space it points to also becomes the interstitial space where they live, away from the prying eyes and limiting socio-cultural factors; they inhabit a place that falls outside and below, literally, the power structures of the dominant patriarchal hierarchies. The courtyard again, even though is located in Lord Indra's domain, is a space outside the confines of the palace and the limitations it may impose upon them. Though the fairies perform at Indira's command and fancy, they still own that particular space. For them, their sisterhood helps them construct their own world, and whenever an attempt is made to breach this space by men, they retreat and think of ways to tackle the situation.

This draws stronger female characters, such as the fairies and the Queen, who show agency and achieve liberation and emancipation through their planning. Research has established the significance of sisterhood in the theoretical paradigm of feminism (Morgan 1970). The fairies within the folktale exhibit solidarity, a sense of collective identity as well as a collective understanding of purpose when dealing with men and unity. Holding a position of power over the young prince, they make him a *tabalchee,* act in a certain manner, and ask for the boon. While seemingly the boy takes revenge, the fairies are the ones who do so through him. Theirs is a transactory relationship with the other gender, as well as a relationship which portrays proximity and control (even if limited) over men. The prince had his way only when the fairies had their own; it was a transaction in which both sides benefited from one another. The performance is a mere spectacle when performed by the fairies for men; it is conducted through their own will and delivers their agency to perform as and when they like it, produced by the solidarity that exists because of the sisterhood between these fairies. The narrative's closure can be achieved successfully owing to the unity and collective identity the fairies show. They show the non-hierarchical relationship emblematic of sisterhood (Henry 2004). The folktale dismisses hierarchical relationships and emphasises parity and equality between the fairies for not much information is given about these fairies regarding their age or relationship with each other (mother, sister, daughter, etc.). In a way, they project a social order that discourages power, control, and hierarchy within the same group. As culturally embedded texts, such narratives set out a moral lesson which further promotes and disseminates harmonious and non-hierarchical relationships between women. The narrative teaches the readers the importance of female friendships and sisterhood, where kindred women can dismantle social hierarchy and bring sustainable socio-cultural parity within society.

Conclusion

Oral narratives help construct an alternate space that leads to the emergence of stronger female characters within the typical hierarchical patriarchal spaces. These female characters become instrumental in rewriting gender norms and dismissing prejudiced ideas concerning female friendships and relationships. This chapter argues that close readings of narratives show sisterhood as an important relationship within the gendered spaces; they are marked by their non-hierarchical nature, which empowers them as well. While there is a need to retell the tales from women's perspectives, the existing ones do offer spaces where those power hierarchies are dismantled in a covert manner. Such oral narratives facilitate the dismantling of gender bias along with helping women reclaim and recreate their identities through a dependence on other women.

Amy Lee

Onmyōji (Yumemakura Baku, 2003)

The Modernisation of Japanese Folklore as Self-Empowerment

The Heian era (794–1185) was one of the most glorious periods in Japanese cultural history. The almost four centuries of relative political stability allowed the rich growth of art and culture, especially in the literary breakthrough and the emergence of female authors. One thousand years later, this period is looked back with fondness and reverence. Contemporary popular fiction writer Yumemakura Baku re-creates the Heian period in his *Onmyōji* series for the 20th- and 21st-century readers through re-refashioning Abe no Seimei and his regiment of spiritual beings. The legendary Heian *onmyōji* re-appears not only with his first-rate ability to manipulate spiritual beings such as the *shikigami*, but also in a new detective image whose process of decoding touches the most profound human psychology. The Abe no Seimei in contemporary popular fiction is a mysterious figure who escapes the usual identity classification and whose power is most visibly manifested in his being at one with nature. The proposed chapter examines episodes from the *Onmyōji* collection to illustrate this aspiration of the new individual self, a focused being who is at one with nature and who possesses the most intimate knowledge of how the human mind works. Yumemakura Baku, through focusing on the ambiguities inherent in Abe no Seimei's identity and highlighting his regiment of female *shikigami* (which are different from the traditional "small ghosts" as they appear in Japanese folklore), has depicted for his 21st-century readers the most relevant abilities to live a successful life in our much-changed world. The chapter argues that Yumemakura Baku is consciously modernising Japanese folklore in his works through the psychological-detection tasks performed by Abe no Seimei (921–1005), and much of his achievements came from a transgression of boundaries, including his regiment of female *shikigamis*.

Stories as Documents of Values

It has often been said that the best way to learn about a people is through examining their contemporary popular culture. Unlike news reports or other official documents which keep a record of events that happen during a specific time period, objects of popular culture reflect the thoughts, beliefs, and feelings of the people at the time. Ancient literary and artistic works are precious records of the roots of many of our beliefs and values today, and very often the changes of such beliefs and values can be tracked in the changes of literary and artistic stories across different eras.

In this chapter, one fairy-like figure with a long history in Japanese literary texts is examined as a mark of changes to how the self is perceived and represented. The *shikigami*, also known as *shikishin* (Yumemakura 2003: 17), is "commonly identified as a spirit servant controlled by the practitioners of Onmyōdō, the *onmyōji* (yin–yang master)" (Pang 2013: 100) in today's popular culture. Western scholars who have shown an interest in *shikigami* have variously used terms such as "spirit", "genie", or "familiar", to describe this being. To understand whether the *shikigami* is indeed the Japanese version of the genie, Pang attempts to trace its original representation in ancient Japanese literature. She finds that it was "an amorphous augury tool" (Pang 2013: 100), and its characteristics to be "rather eclectic and encompassing in nature" (Pang 2013: 100). Finally, having followed through its various representations in pre-modern literature, Pang offers a summary of the five main meanings of *shikigami*: a metaphorical reference to an *onmyōji*'s prognostic powers; a representation of human will and consciousness; a type of inherent energy in objects that can be utilised through spells; a human-created magical curse; and a spirit servant which is also the common understanding today (Pang 2013).

This examination of the *shikigami* as depicted in popular culture today is not to continue Pang's investigation of its religious meanings but to try to understand the contemporary mind through the modernised *shikigami*. Despite the fact that *shikigami* has not been found in any historical records, its persistence from ancient folklore to contemporary popular fiction shows that it has a place in the collective consciousness across different periods. In the following discussion, an example is taken from Yumemakura Baku's hugely popular fiction series entitled *Onmyōji*, to see what the *shikigami* may mean to contemporary readers. The very first short story of this series, "Genjo to iu biwa oni ni toraruru koto" (A biwa called Genjo is stolen by an *oni*), will be used to illustrate the content of the modern *shikigami*, which I argue is an individual character that embodies values easily identifiable to contemporary readers in Japan and the Asian region where the series is hugely popular.

Shikigami: Measurement of the Skills of an *Onmyōji*

While Pang cannot find historical records of *shikigami* in religious texts, there areseveral references to this spiritual being in literary texts of the Heian period (794–1185). They are usually depicted as spirit servants of *onmyōji*s, whose level of competence can be seen from the number of *shikigami* he possesses, as well as the complexity of tasks he commands these beings to do. Although there is a lack of historical evidence that *onmyōji*s have the ability to command *shikigami*s by their magic, the presence of such descriptions in a range of literature "suggests that the people of the time considered that *onmyōji* could command *shikigami*" (Shinichi 2003: 93). In *Makura no Sōshi* (The Pillow Book: 1002), a prose collection written by a court lady Sei Shōnagon (966–1017), we can find a reference to an *onmyōji* accompanied by a youth well-versed in ritual knowledge. Shinichi proposes that "it seems to have been a custom of the time for *onmyōji* to be attended by young boys, and people may have looked on them as *shikigami* commanded by the *onmyōji*" (Shinichi 2003: 93). Indeed, literary descriptions (and even artworks) of *onmyōji*s do feature them as accompanied by young boys.

Abe no Seimei (921–1005), a well-respected *onmyōji* who served in the imperial court during the mid-Heian period, was known to be accomplished at controlling *shikigami* (see Figure 18). Japanese literary classics such as *Konjaku Monogatarishū* (Anthology of Tales from the Past, late Heian period) and *Uji Shūi Monogatari* (Tales Gleaned from Uji, early 13th century) both contain tales about Abe no Seimei commanding *shikigami*s by his magic. Although these are literary depictions, the fame of the historical Abe no Seimei is real, as can be seen in the establishment of the Seimei Jinja by the Emperor Ichijou in 1007, after Seimei's death. The Kyoto-located shrine was destroyed during the 15th century, later rebuilt in the same location to mark Abe no Seimei's residence at the time (see Figure 19). Today the Seimei Jinja is a well-attended tourist attraction, featuring statues and portraits of this Heian *onmyōji* and his *shikigami*, as well as the fictional characters in the popular fiction series.

Figure 18. Image of Abe no Seimei and his *shikigami*, artist unknown, 14th century. Image in the public domain.

Onmyōji (Yumemakura Baku, 2003)

Figure 19. A stone statue of *shikigami* at the Seimei Shrine, Kyoto.
Image in the public domain.

Abe no Seimei Refashioned: Yumemakura Baku's Transgressive *Onmyōji*

Both Abe no Seimei and his *shikigami* gain a new life in the contemporary popular fiction written by Yumemakura Baku. The first *Onmyōji* short story was published in 1986 in a literary magazine and later serialised due to its huge following. In this series, Yumemakura refashions the Heian court *onmyōji* into an unconventional character who resonates with contemporary readers residing in a mainly urban setting. He describes Abe no Seimei as a man who is "like a night cloud that is suspended in the air, flowing with the wind" (Yumemakura 2003: 11). He is extremely talented, sure of his own abilities, and although highly regarded by courtiers, he is also cynical about the social hierarchy and tends to act as he likes, much like a free spirit. His best friend and partner-in-adventure is the courtier and musician Minamoto no Hiromasa, who brings him unsolved mysteries and often accompanies him in the process of righting the wrong. The two of them are paired up like an ancient Japanese version of Holmes and Watson.

One indication of Abe no Seimei's power can be seen in the author's reference to the *Anthology of Tales from the Past*, which tells the story of an old man coming to visit Seimei, claiming his intention to train under him. At this time, Seimei is already an *onmyōji* serving the imperial court and has acquired a reputation. This old man brings with him two teenage servant boys who are actually *shikigami*s under his command. Seimei is quick to see through this plot as merely a test of his powers. Here, the author includes a brief description of *shikigami* from literary classics as a kind of "genie that is normally invisible" (Yumemakura 2003: 17), once again to maintain the sense of historical authenticity despite the fictional nature of the story. Seimei sees that the old man is already quite a competent *onmyōji* to be able to control the two *shikigami*s. He declines the invitation to teach the old man, but to demonstrate his own abilities, he takes over the control of the two teenagers and hides them from the old man. Realising the truth, the old man admits, "I have never seen anyone who can hide other *onmyōji*'s *shikigami*s. I can see that you possess much superior power" (Yumemakura 2003: 20).

Seimei's *Shikigami*: A Gendered Modern Individual

And what kind of *shikigami* does this free spirit of an *onmyōji* choose to be around him? At the beginning of the first story, we are introduced to the *shikigami*s "at work" in Seimei's own residence. The narrator tells us that even in an empty house, the windows will close on their own, and the main door will also shut by its own accord (Yumemakura 2003: 22). Apparently, Yumemakura's Seimei not only has completely mastered the *shikigami*s but has also established a more personal relationship with them. They are almost like residents in Seimei's house. This new identity of the *shikigami* can be seen in the details of how Minamoto no Hiromasa is welcomed into Seimei's house when he comes to seek his help over the loss of Genjo and the mysterious biwa music in the night.

On arrival, Hiromasa sees an empty house, the courtyard of which is untended, and trees, flowers, and shrubs simply grow all over the place. As he is mumbling to himself not knowing whether Seimei is home, a good-looking young woman of about 20 years old appears, dressed in male attire – "casual wear of young nobleman of the time" (Yumemakura 2003: 26) – to receive him. She claims that her master knows of his arrival beforehand, and she is sent to bring him into the house. While Hiromasa is still wondering how Seimei knows of his unannounced visit, they reach the sitting room and the female servant who is dressed in male clothing retreats. Seimei is obviously happy to see his best friend and he calls for wine. As he claps his hands, another good-looking young woman, also dressed in casual wear of young nobleman of the time, comes to serve wine to them. Hiromasa stares at this new face and actually asks Seimei if she is human (Yumemakura 2003: 29).

Knowing Seimei's abilities and habits, probably Hiromasa has a good idea that these good-looking women in male attire are *shikigami*s. To his question, Seimei smiles and replies that he can see for himself when he sends these young women to him in the night. Apparently, this kind of teasing between Seimei and Hiromasa occurs very frequently – Hiromasa being such a simple-minded gentleman who takes everything seriously. For him, visiting Seimei is a very confusing experience, because, in his words, "every time I see new faces", to the extent that he is "confused as to how many people are here in this house" (Yumemakura 2003: 30). Here we have the first indication of the nature of Seimei's *shikigami*: they are "fully human", in the sense that they appear in the form of grown-ups and not small boys/minors being sent around at the master's bidding. It is also particularly interesting that we first see the two (female form) *shikigami*s in male clothing, moreover clothes normally worn by noblemen of the time. These differences not only make the *shikigami* more intriguing, more attractive, but also more of an individual character.

This new individuality in the *shikigami* is further clarified by Seimei's explanation of how a spell is cast. He points to a wisteria tree in his courtyard and says that he has given it the name of Mitsumushi. He explains that because of the name, the tree has acquired an identity, and because Seimei is the person who gives the name, a bond has been created between him and the tree, so much so that the tree will "yearn for his return" (Yumemakura 2003: 34), and even delay the wisteria bloom to wait for him. Seimei calls this casting of the name the simplest spell in the world, but essentially this "magic" is the invisible emotional connection among human beings, something that can only come into being between consenting parties. Later we discover that Mitsumushi is another *shikigami* Seimei uses, a reliable and trustworthy team member when Seimei goes about his business of solving supernatural problems.

The main event of the episode is the theft of a biwa called Genjo, one of the emperor's treasures, five nights ago. People are wondering whether it has been taken by the ghost of a poet who died with a grudge against the emperor. But three nights ago, when he passed by the Rajōmon, Hiromasa heard music so miraculous that it could only be coming from Genjo. When he tried to approach, a decomposing human eye dropped from above to warn him off. Feeling that nonhuman power is involved, Hiromasa comes to seek Seimei's help. Finally, the team that goes to Rajōmon to confront the mysterious biwa player includes Hiromasa, Seimei, a blind monk called Semimaru, who is a distinguished musician, and Mitsumushi, appearing as a young lady dressed in purple. She is given the important task of holding the light and leading the way (see Figure 20).

The mysterious biwa player is indeed a ghost – Kandata, from India, 128 years ago. He has a lonely life, and chance brings him to Japan. He recognises Genjo at the royal palace as the very same biwa that he has made in his previous life. He also sees a court lady who looks like his previous wife, whom he misses badly. Seimei is sympathetic, and they strike a deal: Seimei is to persuade that lady to spend a night with him, and Kandata will return Genjo to its present owner. Seeing that the matter will soon be satisfactorily resolved, Seimei requests Kandata to play some music again. He agrees, and Mitsumushi, who has been witnessing the whole exchange tranquilly, couches, lays down the fire, and dances beautifully to the music (Yumemakura 2003: 65). Her movements synchronise with the music so well that even Kandata praises her dancing.

This scene is full of interesting revelations. Mitsumushi the *shikigami* is trusted with the important task of bringing the team to meet the biwa player and shows her as more than the "small boy" *shikigami* of other *onmyōjis*. However, her self-initiated dancing puts her on an even higher level, for music and dance are art forms that require not just technical skills but also passion

Onmyōji (Yumemakura Baku, 2003)

Figure 20. The image of shikigami in Yumemakura Baku's fiction series *Onmyoji*, adapted in manga by Reiko Okano for Hakusensha (1999). Image used under Fair Use.

and love, which are usually considered to be unique human qualities. Although Kandata later attacks her and turns her back into a bunch of wisteria, for the duration of the dance, Mitsumushi has been elevated to the order of the human being, moreover, an artistic lady who has her own interpretation of the music. At the end of this story, the narrator remarks that the incident is taken from *Anthology of Tales from the Past* chapter 24, reminding readers of its "historical" reality, but at the same time highlighting the very different sense of identity that the characters have acquired in this refashioning.

Conclusion

The *shikigami* is not found in any historical record of the Heian period, but the scattered references in Heian literature show that this figure is present in the collective consciousness of people of the time. This character which is a spirit servant of the *onmyōji* was originally just a tool to show the abilities and sophistication level of the master and appeared as small figures or as children. But in the re-fashioning of the *onmyōji* character in the 20th- and 21st-century popular fiction, the identity of the *shikigami* becomes more visible, and more independent. It is no longer just a small character, neutral in gender and a faceless servant, but an individual capable of their own thinking and decisions and has acquired a distinctly gendered identity. The female *shikigami*s are daring, loyal, and reliable companions to Seimei, and his relationship with them is also personal and loving, more like companions than master and servants. This elevation of the *shikigami* into distinct individuals offers a more inclusive language to describe the possibilities of the human world and opens up the various spaces in this world beyond the boundaries of differences.

Margaryta Golovchenko

Tecna (Iginio Straffi, 2004–Present)

Tecna's Conspicuous Place in the *Winx Club* Franchise

Netflix's live-action adaptation of the iconic Italian animated show *Winx Club* (2004–present) was short-lived. Lasting only two seasons (2021–2022), *Fate: The Winx Saga* was heavily criticised for making several significant changes to the source material. In addition to objections that *Fate* was much more sexualised and explicit in its depictions of substance use and defaulting to angsty teenage tropes (Healy 2021; Kaderabek 2021), which can be explained as Netflix attempting to cater to a teenage audience and perhaps even the Millennials who grew up watching the original show, others (see Oladele 2021 and Paterek 2021) have pointed to two major casting decisions that change the tone of the adaptation. The fairies Terra (the replacement for Flora from the animated show) and Musa, coded as Latina and East Asian, respectively, were portrayed by Eliot Salt, a white actress, and Elisha Applebaum, who is part Singaporean and part European. Both casting changes were met with accusations of whitewashing, a criticism Netflix appeared to take into consideration in Season 2, in which viewers are introduced to Terra's cousin Flora, an earth fairy played by the Latina actress Paulina Chávez. These alterations coexist in addition to an another major change: the omission of the character Tecna from *Fate*.

In the animated *Winx*, Tecna is the fairy of technology. One of the six (after the addition of Layla/Aisha in Season 2) members of the group, Tecna is frequently depicted as the problem solver, using her powers to search for information and to help the Winx get themselves out of tight situations, in as well as participating in various battles against enemies. Although she is initially introduced as the stereotypical logic-over-emotions kind of girl,

Tecna comes to accept her own feelings, both of friendship towards the other Winx and her romantic affections for Timmy, one of the Red Fountain Specialists and her romantic interest throughout the series. In the animated show, Tecna's connection to technology is both literal and magical, extending from her fingertips as much as it is contained within the palm of her hand in the form of various gadgets. Tecna's attacks are rarely named, but the two she does shout – Static Sphere and Plasma Sphere – embody the early 2000s fascination with the internet as the peak of technological development, as something fluid and unseen but that nonetheless connects the entire world in a spherical network. These spheres become a play on Marshall McLuhan's concept of the global village, as the function of connectivity across vast distances must contend with the visualization of weblike connections. When it comes to using technology as an aid, Tecna employs a variety of scanners and holographs that allow her to simultaneously scan the surrounding world while also tapping into an unseen network she regularly refers to in order to obtain more information that could help track an enemy or safely sneak into a location.

Tecna's omission from the Netflix reboot can be read in relation to *Fate*'s ambiguous relationship with fairy iconography and technology specifically. *Fate* is not the first show that appears to understand the concept of a modernised adaptation to mean darker and edgier. *Riverdale* (2017–2023) and *The Chilling Adventures of Sabrina* (2018–2020) are also stark examples of this ongoing trend. Furthermore, some contemporary reimaginings of narratives featuring fantastical beings have chosen to mute the otherworldly, and at times even extreme, elements that defined these very creatures and were previously considered so appealing because of their impossibility. The *Twilight* franchise is one such case; moving away from the campiness of the original *Dracula*, vampires are shown to have chosen a mundane life, rejecting human blood and living peacefully within human society. Similarly, Tecna's absence from *Fate* points to this shaky boundary between fantasy and reality, magic and high-powered technology. It suggests a sense of unease when it comes to inserting magic into a world where technology is ever-evolving but simultaneously deeply embedded into our daily lives, to the point where it is questionable whether it can elicit the kind of shock and amazement that magic is thought to be capable of.

This is not to say that technology is completely absent from the *Fate* series. The Netflix show draws on the narrative device popularised by BBC's *Sherlock*, in which key text messages sent and received by characters in the show are displayed on the screen for the viewer to see. The proliferation of social media, to the point where there is even an off-brand version of Instagram shown a few times throughout the two Seasons, is coupled with the fact that narrative elements one might expect would be fantastical or fairytale-like have been replaced with familiar phenomena from reality. Queen Luna's army drives armoured trucks as they chase the Burned Ones in Season 1, while Luna herself arrives in a sleek luxury SUV in Episode 4. In Season 2, Sebastian runs a store called First World Goods, filled with old electronics like radios and computer monitors used decades earlier on Earth. In Episode 3, there is even a parody of Jeff Bezos in the character of Duke Hammerström, a mind fairy and alumni of Alfea who used his powers to create a delivery service through magic. A more extreme version of Amazon, fairy users of Hammerström's service can make purchases with their minds, their mere thoughts enough to summon that which they desire. Despite this intrusion of the modern world and modern concerns, other elements of *Fate* feel more traditional, closer to the fairy archetype from bedtime stories. In the first episode, Farah (the equivalent of *Winx Club*'s Faragonda) tells Bloom that fairies have not had wings for centuries. After transformation magic had been lost, the fairies in the *Fate* universe evolved to exist without wings. The fairies of *Fate* are shaped more by some semblance of logic and structure than by any unruly magical energy of the universe, to the point where fairy magic is said to be controlled by emotion – the stronger, the more powerful.

It would be misleading to say that Tecna was always out of place among the Winx because technology is incompatible with fairies. Rather, Tecna is the most recent descendent in a cultural lineage of fairies who were in some way connected to technology. Tinker Bell, perhaps the most famous fairy of all (she is even acknowledged in *Fate*, with Farah teasingly telling Bloom that

Tinker Bell was an air fairy), was depicted as a light in the 1904 production of J. M. Barrie's *Peter Pan*. This decision is at once practical, conveying Tink's ethereal quality, and laden with social and scientific significance. In her discussion of Tink's depictions and character, Murray Pomerance notes that the choice to represent the fairy as a light was likely connected to Galvanism, a movement stemming from Luigi Galvani's idea that the body, whether human or non, was a source of electricity (Pomerance 2009: 15). Pomerance goes on to note that Tinker Bell was at once the "fairy of electricity" and "a working sprite", encapsulating electricity's role as both a luxury and "as a utility, a servant" in the human household (2009: 17).

Light's function as a beneficial service was brought to the forefront in the Disney Fairies franchise, which launched in 2005 with a variety of chapter books and an illustrated children's novel. The franchise introduced fairies of different "talents", skills that fairies declare after arriving at Pixie Hollow from the human world to gain a "career" of sorts. Some of the talents that feature prominently in the franchise include tinkering, gardening, animals, water, and light. Fira, the light-talent fairy in the original *Pixie Hollow* chapter book series, was depicted corralling fireflies, who were suggested to have dog-like personalities, while her home contains various sparkling objects like jewels and lamps. The latter prove significant within the Pixie Hollow world, as the implication is that they are powered by fireflies that hide themselves within either flowers or specially designed objects that recall familiar human lamps. The productions of *Peter Pan* and the Disney Fairies franchise both point to technology's complex placement between organic and inorganic, suggesting that it has a natural source but that its representation must always be mediated through some human-made (or fairy-made) form.

Rewatching the original *Winx Club* now, it is difficult not to become more aware of how significant colour is in relation to Tecna's powers. Any spell cast by Tecna comes out green, but not the green that one might associate with the powers of an earth fairy like Flora. Rather, Tecna's green is "Matrix green", a shade of lime green whose acidic quality is hard to mistake for the warmer greens associated with Flora. A recently published interview with Simon Whiteley, the production and concept designer for *The Matrix* trilogy

(1999–2003), reveals that the code was "made out of Japanese sushi recipes", where Whiteley scanned the recipes from his wife's Japanese cookbooks and proceeded to visually experiment with them (Bisset 2017). Episode 8 of Season 2 includes an overt reference to the movie, as Tecna puts on a visor that lets her see the software she coded in the form of a digital city (see Figure 21). In his discussion of *The Matrix* and the film's exploration of what is real and what is not, P. Chad Barnett writes of the difficulties of mapping cyberspace, noting an element of intangibility that masks itself with the false sense of a fixed form (2000: 367–8). Significantly, *Winx Club* flattens cyberspace into a stable visual language that Tecna controls, separating the physical objects connected to cyberspace from cyberspace itself.

Additionally, the animated *Winx Club* appears hesitant in deciding what a fairy should look like and whether the image of the Fae Folk made popular in Victorian England must remain the underlying default. Significantly, John Holmes points to the British Pre-Raphaelite sculptor Thomas Woolner's statuette *Puck* (1845–1847) as the origin for the now-iconic pointed ear shape associated with pixies and elves. Woolner noticed that both humans and monkeys had "a small, blunt protrusion on the fold of the outer ear as it curves over to form the shell", using this scientific observation to depict a purely fantastical being (Holmes 2018: 20). Woolner's iconography was so influential that the ear form came to be referred to as the "Woolner tip" by Darwin and mentioned in discussions of evolutionary theory. Significantly, Tecna undergoes her own Darwinian "evolution" throughout the first four seasons of the show, which formed the Original Series before the show was revived in 2011 with Season 5. In the first two seasons, Tecna's fairy form consists of a sparkling lilac full-body suit, while her "wings" resemble a paper aeroplane constructed out of Matrix-green lines (see Figure 22). Her headpiece recalls the head of the opera singer Diva Plavalaguna from Luc Besson's cult sci-fi film *Fifth Element* (1997), further solidifying the connection to scientific popular culture of the late 1990s.

Tecna's fluctuating appearance takes on a more classical fairy form once she unlocks her Charmix transformation in Season 3, Episode 13. While her transformation sequence preserves the elements of cyberspace transformation

Figure 21. Augmented reality city of code Tecna creates. *Winx Club*, created by Iginio Straffi (Nickelodeon 2005).

Figure 22. Tecna's original form. *Winx Club*, created by Iginio Straffi (Nickelodeon 2004).

also seen in the sequence for her original form (see Figure 23), her Enchantix form has a more revealing dress with dramatic cutouts on the sides. Her wings become a more geometric version of the pointy and upturned wings associated with the fae folk across media, from Victorian works like Edwin Landseer's *Titania and Bottom* (1851) and Joseph Nel Paton's *The Quarrel of Oberon and Titania* (1849) to the Flower Fairies series of Cicely Mary Barker and more recent work of illustrators like Shirley Barber and Brian Froud. More importantly, Tecna's wings in Enchantix form are no longer the odd ones out among the Winx. Tecna's Believix form (see Figure 24) in Season 4 brings back the futuristic costume that covers more of her body than the previous one. Yet, her wings develop in the opposite direction, solidifying their visual resemblance to butterfly wings.

In their critical engagement with the *Winx Club* animated series as a manifestation of Italian neo-liberal values, Nicoletta Marini-Maio and Ellen Nerenberg describe the transformations of the Winx, which they connect to the archetype of transformations within the magical girl genre, as being

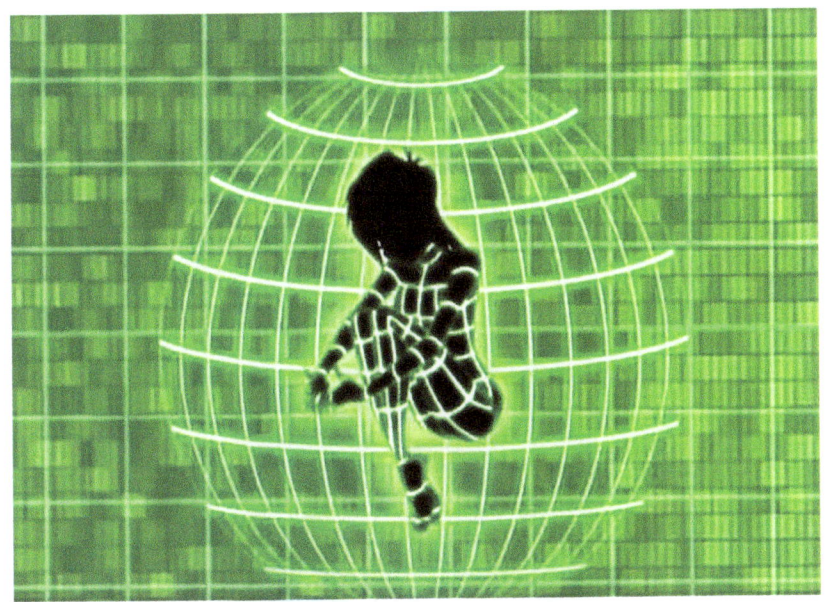

Figure 23. Part of Tecna's transformation sequence for her Enchantix form. *Winx Club*, created by Iginio Straffi (Nickelodeon 2007).

Figure 24. Tecna's Believix form. *Winx Club*, created by Iginio Straffi (Nickelodeon 2009).

simultaneously a form of mutation. The goal of the transformation is, in part, to satisfy the desire to participate in contemporary culture and fit into the landscape of values and aesthetics, from the importance of collective feminine power to the simultaneous awareness of beauty standards and adhering to the latest fashion trends (Marini-Maio and Nerenberg 2020: 29–30). Additionally, Marini-Maio and Nerenberg read the Winx's transformation in conversation with Christianity and the archetype of the pure and angelic woman, as embodied by female saints and the Virgin Mary (2020: 30). In this case, Tecna's

Figure 25. Closeup of magazine advertisement for *Winx Club* reboot. From @winxclub.believix Instagram, 12 July 2023.

Tecna (Iginio Straffi, 2004–Present)

iconographic shift within the animated *Winx Club* as well as her absence from the Netflix reboot may be read in relation to the growing anxiety around the internet and its influence on young girls' perceptions of their bodies. The spread of the internet and fast-paced technology like smartphones (notably, the first-generation iPod touch, the precursor to today's iPhone, was released in September of 2007, months after Season 3 of the animated show had aired) may account for the lack of development in Tecna's powers and even the seeming regression towards a more style-conscious version of the archetypal fairy. In the case of the Netflix show, Tecna's absence may be interpreted to be more of an indicator of the negative connotations that technology, and cyberspace in particular, have acquired in connection with the self-esteem and self-image of teenage girls and young women because of social media. Not only does everyone now carry the kind of portable device that Tecna regularly uses throughout the four seasons of the *Winx Club*, but her place in Netflix's Alfea would eradicate the rigid boundary between the Magic Dimensions and the human world.

It remains to be seen what Tecna's role will be in the rebooted CGI version of the *Winx Club*, which was announced by Ignio Straffi in 2022 for 2024/25. The preliminary promotional materials depict Tecna in a more updated, punkish way reminiscent of the galaxy-tights-wearing Tumblr-era girl (see Figure 25). It seems clear, however, that Tecna's existence is directly influenced by our cultural perception of and relationship with technology, as its pervasiveness down on every level of daily life has removed the mystery and magical quality it had back in the 2000s.

Part VI

Across Media

Butterfly Magic

Kirstin A. Mills

Allyson Wierenga

"The Sums That Came Right" (Edith Nesbit, 1901)

Math is not a subject usually associated with fantasy and magic. However, Edwardian children's author Edith Nesbit joins the two in the form of the Arithmetic Fairy in her story "The Sums That Came Right". Found in *Nine Unlikely Tales* (1901), a collection of short stories, "The Sums That Came Right" tells the story of a boy named Edwin who, after being visited by the Arithmetic Fairy at school, sees the answers to his solved math problems come magically alive. Known for her tendency to merge fantasy with the "tokens of her own time", such as elevators, cars, and aeroplanes (Nikolajeva 1987: 33), Nesbit's choice to combine math with fairies aligns with her narrative style and canon. Furthermore, the Arithmetic Fairy is emblematic of fairies in other works of Edwardian children's literature, such as J. M. Barrie's *Peter and Wendy* (1911) and Frances Hodgson Burnett's *A Little Princess* (1905). In this chapter, I examine the role of Nesbit's Arithmetic Fairy in "The Sums That Came Right" to demonstrate how Edwardian fairies are often constructed as catalysts for childhood agency. In particular, I posit that the Arithmetic Fairy helps Edwin exert agency over his education and magic itself by encouraging him to combine learning with imagination and promoting collaboration between students and teachers. Through this depiction, readers can also see how Nesbit used this story to translate her own theories of education into a fictional narrative.

Though rapid industrialisation during the Victorian age led to increased use of child labour, the flourishing of commercial markets in the Edwardian period coincided with a push for more child protection and labour laws. Not only this, but as Adrienne E. Gavin and Andrew F. Humphries point out, "[c]hildhood in the Edwardian period was a subject of deep concern, fascination, and even obsession" (2009: 1). Edwardian society's deep interest in childhood naturally also extended to literature. Indeed, "Edwardian novels and short stories focused on children to an extent not before seen, nor continued in the same way after the outbreak of World War I" (Gavin and Humphries 2009: 1).

The Edwardians' obsession with childhood is especially pronounced in the unprecedented amount of power that child literary figures have, especially those in Edwardian children's books. Somi Ahn notes that "[b]y presenting child characters who are not in need of adults' help, Edwardian children's texts establish childhood as a fantasy world that seems to differ from the modern secular world run by adults" (2020: 352). In essence, Ahn argues that, as children in Edwardian texts do not need the help of adults, childhood becomes a space free from adult governance. Similarly, Adrienne E. Gavin and Andrew F. Humphries discuss how, unlike Romantic and Victorian writing in which "parents, teachers, employers, and religious figures have enormous control over children", in Edwardian fiction "this adult control dissipate[s]" and "the scales swing triumphantly towards a child power base" (2009: 11).

Published during the Edwardian age, Edith Nesbit's children's texts portray such divisions in power. In particular, Nesbit's children experience power by having access to magic that adults do not have. For example, Paul March-Russell describes how "[f]or Nesbit, magic is a child's prerogative not to be entrusted to adults who, in her novels, remain ignorant [of magic]" (2009: 34). Similarly, Sojin Park states that, in Nesbit's fantasy texts, "It is the children who go on magic adventures, while the British adults are scarcely involved in such adventures" (2010: 514). Nesbit's depictions of powerful children likely connect to her own child advocacy work. In addition to "produc[ing] more than 60 books for juveniles" throughout her lifetime ("E. Nesbit": 2023), Nesbit was incredibly active in the political and social conversations of her day. With her first husband, Hubert Bland, Nesbit co-founded the Fabian Society, a democratic socialist organisation run by "middle class intellectuals" (Smith 1974: 153) whose "primary goal was to 'abolish poverty'" (Cole qtd in Smith 1974: 153). Nesbit also frequently engaged in conversations and events focused on child development, education, and psychology and later published an entire book about her theories of child development titled *Wings and the Child or, The Building of Magic Cities* (1913). As a response to new models of education emerging in her society, Nesbit emphasises a need to cultivate an environment of imaginative freedom and play at home and in the classroom, such as by using fantasy stories like *Cinderella* to inspire learning (1913: 14). By bringing fairies into Edwin's classroom, "The Sums That Came Right" narrativises such notions and adds to depictions of increased child agency in Edwardian children's fiction.

Fairies appearing in Edwardian children's texts frequently contribute to increased child power through their magical abilities and close relationships with child protagonists. Before discussing how this concept manifests in "The Sums That Came Right", I will provide examples from two well-known works of Edwardian children's fiction: Frances Hodgson Burnett's *A Little Princess* and J. M. Barrie's *Peter and Wendy*. In *A Little Princess*, the protagonist Sara Crewe transforms from being a wealthy heiress to an impoverished orphan and servant girl at her boarding school. However, rather than becoming bitter about her change in status, Sara maintains a noble and kind disposition. Sara achieves this attitude by pretending she is a princess, but not just any princess: a *fairy* princess. Midway through the novel, Sara tells Becky, another servant girl, "When things are horrible – just horrible – I think as hard as ever I can of being a princess. I say to myself, 'I am a princess, and I am a fairy one, and because I am a fairy nothing can hurt me or make me uncomfortable'" (1994: 120). In essence, by imagining herself as a fairy princess, Sara can protect herself from pain and discomfort. Presumably, a fairy princess has supernatural or magical abilities, unlike a human one. Such abilities are likely quite attractive for Sara, who is frequently physically and emotionally abused by Miss Minchin, the school's headmistress. As a working-class child completely dependent on Miss Minchin, Sara can do little to physically prevent or control this mistreatment. Yet, by drawing on her imagination to emotionally shield herself and act with kindness, Sara makes her thought processes and behaviour independent of Miss Minchin's control. It is thus largely Sara's imaginative transformation into a fairy that affords her the agency frequently seen in Edwardian literary children.

Unlike Sara, whose fairy-ness is often related to her kind, gentle demeanour, Tinker Bell from J. M. Barrie's *Peter and Wendy* is frequently described as combative, unruly, and mean-spirited. Yet Tinker Bell, who is perhaps the most famous fairy character in Edwardian children's literature, can also be read as a catalyst for childhood agency. When the Darling children – Wendy,

John, and Michael – first meet the mysterious and magical boy Peter Pan, they are mesmerised by his ability to fly. Wishing to fly themselves, Peter Pan tells them, "You just think lovely wonderful thoughts … and they lift you up in the air" (2014: 37). However, readers soon learn Peter was jesting with them since "no one can fly unless the fairy dust has been blown on him" (2014: 37). Having gotten some of Tinker Bell's fairy dust on his hands, Peter blows it on each child, and they instantly have the power to fly. Without Tinker Bell's fairy dust, the Darling children would not have the power to fly and consequently leave their home governed by adults to enter a space run by children alone. Just like Sara's inner strength gleaned from her fairy persona, Tinker Bell's fairy dust leads to increased childhood power.

The beginning of "The Sums That Came Right" begins with a boy named Edwin getting a math sum right and asking, "But what's the use? … Everything else leads to something else, except lessons" (Nesbit 1901: 226). Here, Edwin expresses dismay at not seeing physical proof of his learning. His disappointment leads him to become uninvested in his teacher's lesson and decide to search in his desk for something to play with. Upon opening his desk, Edwin sees a bright light that he assumes to be one of the "firework fusees" he has hidden away (Nesbit 1901: 226). However, the light is actually "the light of pure reason" which "glowed from the glorious eyes of the Arithmetic Fairy" (Nesbit 1901: 226). Although little is known of the Arithmetic Fairy here, that her eyes hold "the light of pure reason" shows her ability to think in a perfectly rational way (as one might expect a master mathematician to do) and creates an instant physical connection between mathematics and reason in her being. Furthermore, the Arithmetic Fairy's mathematical prowess manifests not only internally with her knowledge but externally with her wardrobe. For instance, the narrator describes her dress as "woven of the integral calculus, and trimmed with a dazzling fringe of logarithms" (227). Such a description presents the Arithmetic Fairy as a complicated figure: she is ruled by logic and covered head to toe in mathematical concepts, yet, as a fairy, she is also magical and existing within the imagination and beyond rationality.

In addition to being math personified, Nesbit depicts the Arithmetic Fairy as a teacher figure. In response to Edwin's question about the "use" of lessons, she asks him: "Did no one ever tell you that the things that happen when you've done your sums right, happen when you're grown up?" Boldly, Edwin says, "I don't care what happens then ... I shall be a pirate, or a bushranger, or something" (227). While Edwin believes working as someone outside of society and the law will not require mathematical skill, the Arithmetic Fairy proves him wrong, noting that pirates still need to calculate plunder, and bushrangers must number their bullets. The Arithmetic Fairy thus lectures Edwin on the pervasiveness of math. She teaches him that no matter what role he assumes as an adult, it will benefit from and require math in some way. Her lecture, which includes a reprimand ("Go along with you!" [228]), demonstrates her willingness to assume the role of an older, knowledgeable teacher and push Edwin into the role of a student. In addition to lecturing, the Arithmetic Fairy provides Edwin with concrete instruction. Before leaving, she tells Edwin to "[g]et your Master to set you a little simple multiplication sum in white rabbits" (228). A "sum in white rabbits" is set directly after this conversation, so Edwin clearly respects the Arithmetic Fairy's authority as his educator and follows her instructions.

The Arithmetic Fairy, in true Nesbit style, blends math instruction with imagination. After Edwin's math teacher sets a sum that begins with "seven thousand five hundred and sixty-three white rabbits," the narrator notes that "Edwin ... worked it correctly, by a sort of inspiration, like an ancient prophet or a calculating machine" (229). Here, it seems that Edwin and the Arithmetic Fairy are collaborating to produce the right answer to the sum. It is Edwin who "worked [the sum] correctly", yet the fact that he does so "by a sort of inspiration" suggests that the Arithmetic Fairy's magic is encouraging him. When Edwin arrives at his family's home after school, he finds that "[t]he whole of the front garden, as well as most of the back garden, was a seething mass of white rabbits. Seven thousand five hundred and sixty-three there were, to be exact" (229). In essence, the Arithmetic Fairy, through her magic, has helped Edwin's desire for his lessons "to lead to something else" become fact by making the story problem he answered correctly at school come alive at home. A rapid succession of similar events results as Edwin continues to do his math problems correctly. For instance, the first sum Edwin answers in the story, one about apples, comes to life soon after the rabbit sum does. The narrator describes how Edwin's "cellar was full of apples. Nineteen pounds nineteen and twopence and one-third of a pennyworth – to be exact" (230). At the end of the story, the narrator notes that "The Fellow of Trinity"[1] says the answer to the apple sum is "nineteen pounds, nineteen shillings and two pence and one-third of a penny," the exact price of the apples found in Edwin's cellar (241). And although the narrator leaves the exact answer ambiguous by asking readers if this Fellow "speak[s] the truth" and declaring "I wonder!", it is implied he is correct (241).

As exciting as the living story problems are, they also bring Edwin trouble. For instance, when he correctly answers a sum about "seventy-five pigs travelling in a circle at varying rates", the narrator reveals that "part of this circle ran through Edwin's mother's drawing room" (231). Yet, the narrator says, "No one suspected Edwin of having anything to do with these happenings. And indeed, it was not his fault, so how and why could or should he have owned up to it?" (231). Here, the narrator shows how Edwin and the Arithmetic Fairy are working together again to make these story problems come to life. Edwin, by doing his math sums correctly, "[has something] to do with these happenings." Still, the results are "not his fault" but the fault of the Arithmetic Fairy, whose magic makes the sums become real. The Arithmetic Fairy ultimately instils imagination into Edwin's learning process by bringing his math problems off the page and into reality, and Edwin furthers his education by correctly applying his mathematical knowledge to his schoolwork.

By teaching Edwin about math and the power of imagination, the Arithmetic Fairy provides him with tools to exert power over the come-to-life math problems. Mid-way through the story, she makes a sum about a cistern come alive at Edwin's house. When this cistern causes a huge leak and a deluge like "a sort of Niagara" in his bedroom, Edwin becomes "convinced of the desperate need of finding the Arithmetic Fairy, and begging her to take back the present she had made him" (232). So, he works to figure out how to summon her on his own. Frustratingly, his first few attempts are unsuccessful. For instance, the narrator notes that Edwin "knew there must be something that would fetch the fairy. He said the Multiplication Table up to nine times…. But no fairy appeared" (233). He then repeatedly tries to "fetch the fairy" through applying and chanting different mathematical concepts (233). Even though Edwin's attempts to summon the Arithmetic Fairy at home are unsuccessful,

---

1   Earlier in the story, Nesbit describes this anonymous figure as "a Fellow of Trinity College, Cambridge".

his recitation of many math themes and terms reveals that he has gained both a substantial knowledge of math and the logic of the Arithmetic Fairy's magic, knowledge that empowers him to take control of his situation.

Following his failed attempts, Edwin "quite suddenly ... knew what he had to do. He made up an example for himself" (236). His example is, "If 7,535 fairies were in my desk at school and I subtracted 710 and added 1,006, and the rest flew away in 783 equal gangs, how many would be left over in the desk?" (236). When the answer turns out to be one, Edwin "opened his desk again, and there was the Arithmetic Fairy" (236). Though she chides Edwin for not being "accurate" in his question ("you should have said seven thousand Arithmetic Fairies" [237]), Edwin's math skills are what successfully summon her. He proceeds to ask the Arithmetic Fairy to stop making his math problems come alive and she agrees. By successfully creating and solving a math problem that summons the Arithmetic Fairy, Edwin finally gains power over the magic that has plagued his life for days and discovers that learning need not have tangible results for it to be "real". Edwin has clearly gained this power and knowledge through the help of both his math teacher (who gave him a foundation in arithmetic knowledge) and the Arithmetic Fairy herself. Indeed, the Arithmetic Fairy has shown him how her magic works: if Edwin gets a math sum right, that sum becomes a tangible object. Thus, he knows that if he can work a sum correctly that has the answer of one fairy, then he can likely make the Arithmetic Fairy become tangible to him. Edwin's success echoes Nesbit's ideas about play in *Wings* when she declares, "The prime instinct of a child at play ... is to create ... he will use the whole force of dream and fancy to create something out of nothing" (1913: 9). When Edwin creates and solves a math problem that summons the Arithmetic Fairy out of thin air, he, too, is "creating something out of nothing" via both "dream and fancy," or rather, his imagination.

In his article "Daily Magic", Edward Eager writes, "if there is one thing that makes E. Nesbit's magic books more enchanting than any others ... [i]t is the dailiness of the magic" (1958). The dailiness of magic in Nesbit's "The Sums That Came Right" is certainly enchanting to read about: it is not often one hears of fairies turning up at a school desk and making math problems come to life. However, the *magic of dailiness* in the story is equally enchanting. In "The Sums That Came Right," math, an ordinary subject that students encounter daily in school, becomes magical. The magical quality of math and, by extension, education, is made possible by Edwin's collaboration with the Arithmetic Fairy, a relationship that exhibits the importance of bringing imagination into the classroom as well as supporting child-centred learning. Thus, "The Sums That Came Right" serves as an excellent example of not only how Edwardian fairies serve as catalysts for child agency, but how Nesbit was able to make her abstract educational theories become tangible and accessible to both child and adult readers through the vessel of children's fantasy. And, while schools may not be at the forefront of most of Nesbit's children's fiction, the educational model Nesbit creates in "The Sums That Came Right" – one in which students and teachers work together, and learning is generated in and outside a classroom – can apply to any of Nesbit's other stories, both realistic and fantastical, in which learning takes place. If the Arithmetic Fairy is right (and I believe she is) and there is "a fairy for everything you have to learn", then there must be a fairy for education itself. Is Edith Nesbit that fairy? I wonder!

Péter Kristóf Makai

Changeling: The Lost (White Wolf, 2007)

Our Fairies Are Different

Changeling: The Dreaming (White Wolf 1995) and *Changeling: The Lost* (White Wolf 2007a) are tabletop roleplaying games in the *World of Darkness*, a narrative universe created by White Wolf Publishing for their set of supernatural storytelling systems. United by the fact the players are an informed and persecuted minority of mythical creatures in a world dominated by unknowing humans, they allow transgressive forms of storytelling, focusing on the Gothic, the macabre, and the tragic. Vampires, ghosts, mages, werewolves, golems, and indeed, fairies each get their own storytelling systems to facilitate play that is meaningful to their journeys as kindred spirits against a world from which they have to hide in the shadows. Common to all *World of Darkness* RPGs is a heavily stratified social hierarchy among the denizens of the supernatural world, a focus on social play with intrigue, diplomacy, and backstabbing, a sense of longing or desire that pushes characters into new territories, and a tinge of insanity that haunts the player characters, threatening them with a loss of humanity and therefore the end their in-game agency. Here, I intend to highlight how the fair folk are depicted in the more modern version of the game, *Changeling: The Lost*, arguing that their depiction is a historical evolution responding to changes in social formations in the 1990s, and their presentations of fairies are reimaginations of these magical creatures as we know them.

Although *Dreaming* originally intended to be a less foreboding take on the *World of Darkness*, suffused with the dreamy melancholy of works like Neil Gaiman's *Sandman* comic book series, its mechanical cumbersomeness and its auxiliary nature to the more successful vampiric, sorcery-laden and lupine instalments of the *World of Darkness* meant that it was fit for a darker and more streamlined rebranding. In both versions, players embody supernatural folk (such as the fae, pookas, selkies, sídhe, etc.) who get to live in the primary world of our time, the "mundane" world, and also visit different depths of the Dreaming, the world of fairies. In this dynamic of the Dreaming, the Storyteller (known in other systems as the Game Master) can allow varying degrees of departure from the rules that govern the world of ordinary mortals. The liminal space between the two lands is known as the Hedge, a land of passage that changelings must brave to escape their captors, with its own logic and perils.

The Storyteller system, used by all *World of Darkness* RPGs, was developed "to lessen the importance of props, dice, and rulebooks in gaming, and put more emphasis on narrative and creativity. The game still uses dice rolls to represent success, failure, and luck, but the importance of turning those technical elements into narrative is emphasized" (Zalka 2012), thus ushering in a new generation of RPGs. However, as White et al. point out, even though "[a]t its best, the Storyteller procedures facilitated highly immersive character-focused drama as the PCs confronted the choices laid out for them by the development of the narrative; at its worst, it led to disengaged play as they either resisted or succumbed to the Storyteller's 'railroading'" (2018: 75), that is, the restriction of player agency to proceed with the story as prepared at the cost of collaborative plotting. This owed much of the original system's allegiances to a restrictive vision of what storytelling could be, relying on the Hero's Journey as a template to be followed. In the end, *The Lost* became a more balanced and accessible version of the *Changeling* brand, in line with the rest of the "New World" or "Chronicles of Darkness" universe. This survey does not investigate the mechanics of the game, or changes in editions, but looks at *Changeling: The Lost* (hereafter known as *CtL*) as a narrative setting and explores what aspects of the fairies are preserved from folklore and what traits were adapted to the iconic setting.

As the 2007 rulebook is quick to explain, fairy tales change as audiences mature. Fairies in Andrew Lang's coloured anthologies were tiny, mischievous, and surprisingly powerful supernatural creatures who wreaked havoc on mortals. Lang himself "established principles that over time became normative", and "part of a colonial discourse. With the help of the colour fairy books, Lang effectively colonised fairyland" (Sundmark 2005), and our imagination, too. As *The Lost* sourcebook expresses it, "[t]he good fairies bless the heroes so they can overcome their challenges, and the wicked fairies' curses ultimately come to naught. Everyone lives happily ever after" (White Wolf 2007a: 12), except, it is never that easy. The RPG leans heavily into the more sinister undertones, the overpowering glamour of the fairies, and their forlorn, in-between nature. "Blood and sex creep into the tales. People come to bad ends. [This] is a game about what happens when these old stories prove true" (White Wolf 2007a). Storytellers are explicitly asked to "tell stories of madness, intrigue and dark beauty" (2007a: 245), "or inchoate nightmare" (2007a: 240) to set the mood right.

In *CtL*, players inhabit the roles of a changeling, an ordinary human stolen by the fair folk, "who has been gradually changed by her durance in Arcadia, becoming partly fae herself" (White Wolf 2007a: 16), and they bound together in a *motley*, Changeling-speak for "adventuring party". Each's physical appearance, or *seeming*, is the way that they appear to mortals, while still retaining their *kith*, or their essence (otherwise known as race and class in RPGs). Crucially, they are not True Fae, for they are of mortal origin (capitalisation is a matter of life and death in the realm of Faerie, apparently). In contrast to the changeling stories we grew up with, in *CtL*, "the majority of changelings encountered as adults weren't taken as babies. While many humans are stolen from the cradle, their mortality rate is very high [...]. In fact, no human taken to Faerie as a babe has returned to the mortal world on his own cognizance" (White Wolf 2007a: 23). Faerie, otherwise known as Arcadia, is known for operating under different laws compared to the mundane world, since "[t]he laws of physics and science do not hold sway in Faerie. All reality is based on [...] inordinately powerful Contracts and oaths, and without a Fae mentor to include a human in them, a human's fate in Arcadia is sealed", and as a result "[e]verything in Arcadia exists and interacts as a result of Contracts and oaths with those around it" (2007a: 24). In practice, this means that a concept similar to Celtic *geas* is observed, and is a binding agreement, often becoming the catalyst for quests and entire story arcs in a role-playing campaign.

"The Fae don't fall into categories" (White Wolf 2007a: 65), as the manual claims, and yet, this is not strictly speaking true. Despite the plethora of supernatural entities that fall under the umbrella, changelings are categorised according to seeming, kith and court – cross-pollinating social categories, each with game-mechanical implications. The manual is quite adamant to assert that they are "not so much social groupings as they are a vague, general descriptive shorthand for how different changelings have been changed by their experiences" (White Wolf 2007a: 99), yet these are identity-forming and intersecting social groups, and function as such. In the interest of brevity, I focus only on seeming and kith, although the courts of the four seasons would also merit discussion. As stated above, seeming and kith map onto the RPG terms "race" and "class", but they do so somewhat uneasily. They are taxonomies of predispositions, natural capacities, advantages, talents, and learned abilities that distinguish characters from one another: multifaceted, equal parts restricting and liberating.

The idea behind the seemings is that ordinary people cannot survive their trip to Faerie unscathed. Every contact with the Perilous Realm comes with a price: The Lost were once toys, used and abused by True Fae. Slavery is the word the designers use, and advisedly so: "Abducted without warning and enslaved by almost incomprehensible beings" (White Wolf 2007a: 32). Changelings' freedom is hard-won. They receive their supernatural abilities in exchange for inhuman servitude. They are touched by the magic of the other world, and this is indicated to be harrowingly ambiguous: "They carry their seemings like scars. Seemings are the permanent mark of terrible trauma. At the same time, they're a badge of honor" (White Wolf 2007a: 99). The seemings provided by the manual are "Beast", "Darkling", "Elemental", "Fairest", "Ogre", and "The Wizened". Beyond that, players may (but are not required) to choose a kith as "an optional subcategory" (2007a: 76), such as a "Windwing Beast", or a "Fireheart Elemental", to round out the character.

To begin with, a Beast is a human who was once truly animal but renounced the pure sensory joys of living as one, and returns to the world of humans with an animal's atavisms – not dissimilar to a feral child, who were once thought to be changelings even in our world, and who would be possibly characterised as autistic today (Albury 2011; Renner 2016). They have kiths like "Skitterskulks" (bugs and creepy-crawlies) or "Swimmerskins" (aquatic creatures), or "Venombites" (poisonous/venomous reptiles and insects), largely corresponding to a class of animals. Darklings are ghostly creatures of various stripes, whose kith include the nerdy "Antiquarians", peddlers of occult knowledge, the necromantic "Gravewrights", the shape-shifting "Mirrorskins", or the soul-sucking "Leechfingers", who literally siphon life-force away from humans. Elementals are spirits deriving their powers from natural forces to eldritch ends, including the "Manikins", golem-like Galateas come alive; "Woodbloods", "ye olde fairies", and "Green Men", spirits of the forests; or the "Earthbones", "who have the mark of earth and stone: lumpen Paracelsian Gnomes, sand spirits, dour men of peat and dwarfs made of mountain granite" (White Wolf 2007a: 109). The Fairest are glamorous tempters or temptresses, the celebrities of the changeling world, counting among their ranks the Muses, who inspire great art and ensnare great artists, "Dancers", who are Mata Haris and femme fatales, and the "Flowering", hippies walking in the wake of petals wherever they step. Ogres are a seeming of bruisers that embrace their brutish nature. More racially discriminated against in the game worlds even compared to other changelings, they could be Cyclopeans, crippled and disfigured, Gargantuan, forced into giantry by cruel magic, or "Stonebones", echoing the spirit of Scandinavian folklore's trolls and Native American mountain spirits. The Wizened are broken, emaciated creatures, either short of stature or tall, gaunt, and haggard individuals, and talented craftspeople, such as the "Chatelaines", somewhere between middle managers and majordomos, "Oracles", your fortune-tellers and delphic soothsayers, and "Brewers", whose natural talent lies in creating intoxicating beverages.

As can be clearly seen from even such a short survey, *Changeling: The Lost* runs the gamut of the traditional Tolkienian fantasy race mix of humans, elves, dwarves, halflings, orcs, and trolls and reinterprets them to be more culturally diverse and inclusive. It must be said that despite the designers' will to erase the strong boundaries between race and class as traditionally understood in classic RPGs, the game still adapts a mélange of cultural influences in a way that is mostly catering to a Western audience, with more esoteric changelings fleshed out in supplementary materials, like *Winter Masques* (White Wolf 2007b). Still, the wealth and breadth of seemings and kiths on offer in the base game are already opening up the floor to diverse cultural influences.

In terms of the cultural politics of changelings, they straddle a fine line between being fiercely protective of their identity and welcoming anyone who is non-normative or has gone through the same ordeal they did. And although players running against the stereotypes of each category are encouraged by the rulebook, an interesting tidbit is that *the kiths in-universe stereotypes* are explicitly spelt out on the pages detailing character creation (2007a: 70–122). Due to their long lives, they are said to be more conservative and out of touch with the modern world in some respects, but not in their community-building. One clear indication of this is the sheer amount of lore and obscure terminology one has to absorb when first learning the game. Intentionally designed to be dense, the lexicon shows how secretive and arcane the changelings have become, developing their argot to protect their own.

The freeholds inhabited by changelings are halfway between ghettoes and safe spaces for alterity and their impact is explicitly historicised in the manual: "The modern era and, especially, the Internet have provided a new, tentative outlet for changeling society. Before the late 1980s ... learning about changelings from other cities or other nations relied greatly on reports from wandering changelings" (White Wolf 2007a: 32), resulting in a large, informal, and highly secretive network of acquaintances, in a clear echo of queer folks in the era of criminalised homosexuality and/or the cusp of the HIV crisis. Still, "[b]y the end of the 1990s, there was a loose network of several scores of changelings across the world who had managed to get in contact" and this has allowed them "to travel to different cities and even different nations with some idea of how to make contact with a local freehold" (2007a, 32).

In fact, it is very easy to read changelings as queer-coded, and very hard not to: their alterity has both an innate and a presentational component, a general scepticism of mortals, hiding their true nature, coupled with the fact that revealing their true nature often results in persecution. In modern readings of mythical changeling stories, "[t]he changeling provides a perfect metaphor for the common feelings of alienation and estrangement experienced in youth and also offers a magical resolution to these problems in the realization that difference is due to the child's mystical and special identity" as Renner argues (2016: 172). Even if this is more general in their recognition of teenage angst, although ordinary teens are likely to experience both alienation and estrangement, LBGTQ+ adolescents would feel so doubly. Likewise, mental health issues, although problematically romanticised as beautiful madness, are a focal point of the gameworld and its storytelling. Changelings also pledge to their motley to become their found families, reinforcing their outcast nature and a desire not to be outed. Their freeholds are bastions of "tolerance for both eccentricity and dysfunction" (2007a: 39), the latter being a less fortunate term for presenting as neurodiverse, but its inclusion as part of the changeling identity is a welcome possibility for working through the impact of non-typical development and queer awakening or gender dysphoria. Someone literal-minded enough might even wonder if the game's conception of changelings could be the literalisation of the fact that the slang term "fairies" are used (derogatively) for homosexuals. Nonetheless, since the game was originally published in the 1990s, it is hardly a stretch to observe that these themes would speak to our awareness of LGBTQ+ people and the much-needed advancements of their rights.

Catalina Millán Scheiding

Prikosnovénie (Frédéric Chaplain and Sabine Adélaïde, 1991–Present)

Contemporary Fairies of Old

From the genre-breaking Dead Can Dance, which found a niche in the dark-wave scene with their personal reimagining of world music traditions, to the appearance of Sex Gang Children's frontman Andi Sex Gang, called "the Gothic Goblin" (Carpenter), the dark alternative music scene has frequently included mythological, literary, and folkloric influences at the musical, thematic, and aesthetic levels. Numerous dark folk and neofolk bands have introduced themes from pre-Christian beliefs or from pagan references, including Catharism, Norse, and Greek mythology, and other heathen spiritual movements, through bands such as Hagalaz' Runedance, Narsilion, Sonne Hagal, or Sol Invictus. Some artists have created performing identities which have included depictions of the Fae world, from gnomes to boggarts and goblins, such as Mortiis, Corvus Corax, or Sopor Aeternus while many bands have touched upon fairy-related themes, both by neoclassical or Celtic influenced artists such as Loreena McKennit or Faith and the Muse as well as by more electronic, rock, and metal bands (ASP, Type O Negative, Indochine or Blutengel).

Fairfolk could be considered one of the main inspirations for the Prikosnovénie label and its artists. Prikosnovénie invites the listener to "enter a 'fairy world' universe full of marvelous voices, mythologies, fantasy worlds", as an introduction to their compilation "Fairy World I". While the musical styles Prikosnovénie includes vary from heavenly voices to neoclassical music among the dark-wave scene, the description of these styles picks up the fairy leitmotif with a subdivision into several different labels, such as Fairy Romanticism, Fairy and Elfish Lands, or Electronica & Ethnica. This chapter intends to explore the three modes in which Prikosnoveniénie represents and reimagines fairies as a cultural icon: music, visual art, and live concert, specifically through their "Fairy World" compilation releases (from 2003 to 2009), their 100th release ("Effleurement" 2007) and their music festival "La Nuit des Fees" (The Night of the Fairies) celebrated from 2007 to 2009 in the medieval city of Clisson.

History

The label was created in 1990 by Frédéric Chaplain and Sabine Adélaïde after they met at film school in Nantes; it was created to merge music and imagery and generate a holistic experience of music listening (Chaplain and Adélaïde 2003). Their releases focused on a niche market that incorporated mostly heavenly voices bands, as well as numerous neoclassical artists, and slowly began to incorporate electronic and ethnic productions. Soon after this shift, by the late 1990s, Arnö Pellerin became a co-manager of the label.

While Prikosnovénie initially initially released song collections that were more focused on dark and new wave, they moved towards the direction of folk and heavenly voices styles after the release of their second compilation and the Cherce-Lune full album release. The first "Fairy World" compilation, released in 2003, presented a broad designation that intended to unite different versions of the same idea, from ethereal to pop and electronica. Subsequent compilation releases (2005, 2007, 2008, 2009, and 2010) already presented several sublabels for more niche genres of music, such as solaris, lunäe, supernove, LYTCH, or Mandalia. "Fairy World IV" introduced the idea of "dream sowers": musical imps that germinated the new Prikosnovénie releases, who were "close to the fairy's DNA, incarnates as giant musical flowers. Their fragrances disseminate to feed your body, your physical, sensory organs, offering a wide range of sonic emotions, such as sorrow, nostalgia, melancholy, anger, enthusiasm, ardour, passion" ("Fairy World IV").

The early 21st century was a golden moment for the label's releases, with their own roster of artists, and reissuing and signing some wave and folk-renown groups such as Collection d'Arnell~Andrea, Ataraxia, or Louisa John-Krol. By 2015, the label stopped numbering their releases and, since then, has become a side project of Mandalia Music, focused on folkloric and traditional instruments and their use for relaxation and zen music production.

Visual Work

Aesthetically, Prikosnovénie draws on the influence of different traditions of "small gods" (Robbins 2011) and the conversation between the materiality or "spirituality" of these traditions (Espírito Santo and Blane's 2014). Images depict different creatures in natural environments, but they are always isolated from any specific context.

Visual art for the main compilations of the label has been created by Prikosnovénie's co-founder, Sabine Adélaïde. The art that accompanies the Fairy World compilations uses mixed techniques to display creatures that could be recognised as belonging to different traditions of fae, including elves, leshiye, gwragedd annwn, kobolds, nymphs, fantines, or sirens, among others. While most of the art is hand-painted using pencil and watercolour, techniques include photography, collage, and image superimposition. Fairy World Compilation I includes different "creatures" with each page, while the images that accompany the electronic and ethnic releases are blurred and mostly monochromatic. Upon later releases, figures belonging to other spiritual traditions are included; for example, in Fairy World IV there are at least two images of Eastern religions. This combination of creations follows the inherent indeterminacy that Purkiss acknowledges within the depiction of fae, while underlining the double reading of these images as either symbols of the censored or the unattainable (Purkiss 2001), or real beings that can be interacted with (Ostling 2018: 5), preserving their ambivalent and ambiguous nature (Ostling 2018: 9). Additionally, the depiction that Prikosnovénie offers of fae is mostly related to nature and wilderness, instead of their domestic counterparts (Ostling 2018: 8) with numerous images displaying natural elements, flowers, trees, and leaves (see Figure 26).

Music

Prikosnovénie's Fairy World compilations merge folkloric and contemporary multinational music, from Australia to Japan, including Balkan and Sufi music, Tuvan throat singing, Greek Antiquity, and Middle Eastern influences as well as modern dark wave. As part of the multifaceted concept of Fairy Music that Prikosnovénie offers, label musicians become storytellers of their own perception of fae. Using the Russian band Caprice as an example,

Figure 26. La Feerie/ A fairies world, card #2, illustration by Sabine Adélaïde, Effleurement, Prikosnovenie, 6 November 2006.
Reproduced with the permission of the artist.

their motivation might appear in the making of the music, as shared in an interview:

> I can't say where any music comes from. But we have read a lot of descriptions of real faerie music in several books which (among other faerie-related things) talk about the experiences of people who somehow managed to get into the parallel world (where faeries dwell) and hear their music. Our music is an attempt to show what the music written by elves, not by humans, sounds like (Monteiro 2002).

And it can also be found in the thematic selection, such as the focus on specific stories in the elven lore of Tolkien, with their release of "Of Amroth and Nimrodel" and its "stark other-worldliness" (Sturgis 2010: 128).

On the topic of *The Lord of the Rings* as an inspiration to represent the world of fairies and elves, Shore's work in the film trilogy based on Tolkien's work depicted fae music by the use of non-diatonic modes, such as Dorian, Phrygian, Mixolydian, and Lydian. Only some of the label's artists follow this technique to depict the unfamiliar and the fantastic (Rone 2018). The use of classical and traditional instrumentation is more prevalent and is used to recreate the nostalgia that is commonly linked with faerie representation (Ostling 2018: 12). Daemonia Nymphe introduces traditional Greek instrumentations such as lyre, double flutes, or crotale; Slovenian Pinknruby include the bağlama, and the dulcimer as well as the harp; G.O.R, the project spearheaded by Italian classical musician Francesco Banchini, incorporates chalumeau, oud and ney, among other instruments. Some of the dream pop artists, such as the Ukranian band Fleür, French Collection d'Arnell~Andrea, and English Mable Bee include strings and classical piano to their songs.

Another technique used to represent the otherness of the fairy world includes the mix of genres, with some songs using atmospheric electronic backgrounds under neoclassical instrumentation. Bulgarian band Irfan, included in five of the six Fairy World compilations as well as in "Effleurement", effectively sets an otherworldly mood through this technique.

Themes also mention natural elements or pay homage to fae and folklore. As previously mentioned, Russian band Caprice, included in five of the six Fairy World compilations and in "Effleurement", consider that they write the "music of the elves", inspired by Tolkien themes. Daemonia Nymphe's work is inspired by nymphs, their relation to nature and their reactive nature to interaction with humans (Van Muylem 2013). Artist Louisa John Kroll regularly includes Arthurian characters, fairy tales, and literature in her lyrical songwriting. About the origins of her work, she mentions:

> Singing plants and poems into melody, and believing in dryads, I made a garden with a wetland, frog pond and flowering vines, ripe for solitude. Fantasy chronicles of Earthsea, Middleearth and Narnia inspired me, as did *Faeries* co-written by Alan Lee (who went on to design sets for Tolkien films) and Brian Froud (who later made a music clip "Muse" for my music). (Mason 2018)

A special mention should be given to the gender perspective of the depiction of fae in Prikosnovénie, as it is heavily influenced by and one of the best representatives of the musical sub-genre called heavenly voices. Considered as a type of ethereal wave, the main linking quality of the genre is the use of female vocals. The release of "Effleurement", Prikosnovénie's 100th album, is a prime example of this style, as it offers a collection of 17 songs fronted by women singers or creators. The prominent presence of women might follow "the widespread trope of the alluring female fairy of the forest or the streamside" (Ostling 2018: 26) seen in mythologies worldwide, while it also connects with the Romanticism tradition of fairy painting and the dominant presence of the female fairy (Maas 1997). This connection between women and the "otherness" of fae is echoed by the collection of card images included in the release, which represent different female virtues illustrated by a specific song of the compilation (see Figure 27). The listener is invited to shuffle through the cards, choose an image and read the accompanying text while listening to the indicated song. The texts that accompany each card were written by Frédéric Chaplain.

Live Concerts

In 2007, 2008, and 2009, Prikosnovénie celebrated the festival "La Nuit des Fees" in the medieval halls of the Chapelle St-Jacques of Clisson. The events included some of their signed bands, such as Collection d'Arnell~Andrea

Figure 27. La Face Sombre/The Dark Side, card #9, illustration by Sabine Adélaïde, Effleurement, Prikosnovenie, 6 November 2006.
Reproduced with the permission of the artist.

(2007), Irfan, Ashram, or Antrabata (2008), Daemonia Nymphe, Corde Oblique, or Christa Galli (2009). Female presence was also dominant on the stage, and neoclassical or traditional instrumentation appeared in many of the acts, while some of the more pop-oriented bands performed with strings.

The event mostly picked up medievalist and Anglocentric perceptions of the Fairy World, taking inspiration from Victorian Romanticism and offering services that could be analogous to many of the Renaissance fairs that take place in medieval locations of Europe. Some audience members were dressed as fairies or wore medieval-inspired garments. The festival also offered a marketplace that included art pieces (many of them specifically inspired by the fairy world), illustrations, a literary selection of fantasy books, crafts, and decorations. Additionally, there were courses offered in harmonic singing, cinema, sword fighting displays, and dances, among other activities. Since the location is urban, many of the nature-focused ideas that Prikosnovénie's music displays were represented through a nostalgia for a pre-industrial setting, as found in the tradition of fantasy (Niiler, LeGuin): a 12th-century building, numerous classical and folkloric instruments, a focus on crafts, niche activities.

Conclusions

A brief overview of Prikosnovénie's main work, and a more specific examination of their "Fairy World" compilation releases, "Effleurement" and "La Nuit des Fees" display how the label created their own narrative in relation to the sound and look of contemporary Fairies, while they constructed an effective and immediately recognisable brand in the dark alternative scene. The inclusion of unique musical modes, distinctive instrumentation, and the clear dominance of heavenly voices, combined with topics introduced by the lyrics, represent the integration of tradition and modernity. In the case of Prikosnovénie, both the artists and the imagery are heavily linked to a unique, melancholic, and ingenuous way of connecting to nature. The demonic nature of fairies after Christianisation is disregarded, and there is no conflict in the combination of different spiritual options. Islamic Sufism, Pre-Islamic African folk, Buddhism and Zen, Catholicism and European traditions, Paganism and Wiccan practices find solace in a label that links all of these traditions under a focus on Mother Nature and the small gods of wildlife, earth, and water, through the omnipresent visual work of Sabine Adélaïde. Fairies embody the spirit of nature and a return to crafts and folklore, without shunning the contemporary music options that the modern world offers. On the connection of nature and the fae, Prikosnovénie aligns with fairy narratives serve primarily to reenchant the natural world at a time of unprecedented ecological crisis. They animate and personalize it, creating emotional links between practitioners and places, plants, and animals. They are part of a body of imaginative responses to an environment in crisis that ascribe meaning to it, creating a participatory consciousness that may impel people toward more sustainable practices. (Magliocco 2018: 329)

With nature as a main theme that underlies the musical, the visual, and the performative, Prikosnovénie displays the ability to transcend stereotypical depictions of faerie and reimagine a contemporary space for fae. In her seminal work, Briggs (1978: 7) pointed out how fairies are "always vanishing and always popping back up again". In the case of the French label, the space that they created was one populated with the knowledge of the faeries of old and equipped with the options to create contemporary magic.

Kirstin A. Mills

The Fairy-Land of Science (Arabella Buckley, 1878)

Fairies and science might seem like unlikely companions. Like chalk and cheese, night and day, fairies and science seem to exist at opposite ends of a spectrum: the dreamy, magical realms of imagination and the cold, hard facts of reason. Yet fairies and science live in much closer proximity than we might realise. Fairies are, after all, typically considered invisible forces of nature, and how else might we describe the basic concepts of science, such as gravity, magnetism, and energy? At their surface, both fairies and science involve invisible forces and can offer seemingly strange explanations for even stranger phenomena. Indeed, fairies and science have lived in close proximity throughout much of history, evolving together as humans attempted to distinguish supernatural belief from scientific fact. Even in the Age of Reason and the subsequent Victorian period, when science exploded throughout Britain and Europe, becoming not just a scholarly pursuit but also a popular and fashionable topic of leisure and interest, fairies were right there alongside it, enjoying a renaissance of their own. Fairies were everywhere in the 19th century, accompanied by the era's love of fairy tales, and both, as we shall see, were often brought together in fascinating conversation with science. This conversation went both ways, where stories about fairies often linked them with scientific concepts, often as a way of implying the possibility of their existence in the natural world. Likewise, science education for children often drew on fairies as a metaphor to help explain the invisible forces of the natural world. Since then, the fates of fairies and science have remained so intertwined that even in the 21st century we can see the same use of fairies to communicate science for children, and the same representations of scientific forces in our stories about fairies.

From their earliest days, fairies have been linked with the invisible energies and phenomena of the natural world: with flowers and woodlands, with brooks and frosts, with rainbows and toadstool rings. Fairies are spirits of nature in much folklore of the world. In 19th-century Britain, interest in fairies exploded around the same time as a widespread public interest in other phenomena so exciting it seemed like magic: science. Discoveries in electricity, steam, geology, astronomy, and meteorology, to name but a few fields that seemed to turn the world upside-down, made science seem like a never-ending fairyland of possibility. One of the most interesting products of this parallel rise of interest in science and fairies was a genre of children's books that used fairies and the power of storytelling to communicate scientific concepts in a bid to transform the "bundle of dry facts" comprising traditional science education into something more appealing to the young mind. Foremost among these more imaginative science writers was Arabella Buckley, whose book *The Fairy-Land of Science*, first published in 1878, became a popular and influential text throughout the rest of the century. The central conceit of Buckley's text is that the "real" fairies of the world are the invisible forces, such as light, heat, and chemical attraction, that 19th-century science continued to illuminate as the fabric and processes of the natural world. Buckley's opening lines, in which she contrasts her scientific fairies with the more familiar fairies of literature and art that pervaded the Victorian period, are indicative of her project and its attempts to harness the magic of fairies to ensnare young imaginations in the service of scientific education:

> most of you probably look upon science as a bundle of dry facts, while fairy-land is all that is beautiful, and full of poetry and imagination. But …science is full of beautiful pictures, of real poetry, and of wonder-working fairies; and what is more, I promise you they shall be true fairies, … and though they themselves will always remain invisible, yet you will see their wonderful poet at work everywhere around you (Buckley 1880: 1–2).

The invisible quality shared by both the fairies of poetry and Buckley's "fairies of science" is key to the metaphor's power: after all, who are we to say that these are not the same invisible "*forces* around us, and among us", at work (1880: 6)? It is also key to the readers' understanding of science, where we can often only observe the impact of these forces upon matter, and not the forces directly. Yet, wary of disappointing the reader seeking the flitting fairies of

The Fairy-Land of Science (Arabella Buckley, 1878)

poetry (see Figure 28), Buckley assures us that the fairies of science "are ten thousand times more wonderful, more magical, and more beautiful in their work, than those of the old fairy tales" (6).

Indeed, "science tells" its own "fairy tale of nature" (Buckley 1880: 4) and Buckley's book aims to help child readers "make the acquaintance of the science-fairies themselves" (5). Over her following chapters, Buckley introduces a range of different "invisible fairies of nature" (8) explaining not only the scientific principles behind the natural phenomena but also how readers might perform their own scientific experiments to observe them in action. Among the quite extensive list of such fairies are: "the hidden fairy 'life'", responsible for the growth of living things such as budding plants, flowers and trees (10); the "strange fairy, 'Electricity', [who] flings the lightning across the sky and causes the rumbling thunder" (12); the fairies of "*gravitation*" and "heat" (8, emphasis in original); and even "the invisible waves which make our sunbeams, are wonderful fairy messengers as they travel eternally and unceasingly across space, never resting, never tiring in doing the work of our world" (48). We encounter too the "wonderful fairy power" called "*chemical attraction*" (16, emphasis in original), which keeps atoms bound in molecules, and "up in the

Figure 28. "Dancing Fairies" by August Malmström, 1866. Image in the public domain.

clouds" we meet "another of our invisible fairies, which, for want of a better name, we call the 'force of crystallization,'" and which works with the fairy "cohesion" to mould "tiny particles of water [...] into those delicate crystal stars known as 'snowflakes'" (10). Exhorting her readers to adopt scientific curiosity about the natural world, which they might satisfy through observation and experimentation, Buckley reminds her readers: "We have only to stretch out our hand and touch" these natural phenomena "with the wand of inquiry, and they will answer us and reveal the fairy forces which guide and govern them" (236).

Buckley's book was not the only one to draw on the "enchantments" of the fairies (Buckley 1880: 3) to explain science: as Melanie Keene explains, the work typifies a vast genre of such literature emerging in the Victorian period (2015: 18–19). Even literary fairy tales, such as Charles Kingsley's influential *The Water Babies* (1863), drew an explicit link between fairies and the sciences of the natural world. Kingsley's story revolves around a boy transformed by "the fairies" into a water baby (a diminutive, magical creature, often mistaken throughout the text for a newt), and his subsequent adventures through the strange realms of nature. To justify the possibility of such creatures as fairies and water babies, Kingsley draws on science:

> There is life in you, [...] which makes you grow, and move, and think: and yet you can't see it. And there is steam in a steam-engine; and that is what makes it move: and yet you can't see it; and so there may be fairies in the world, and they may be just what makes the world go round (1994: 40).

Such arguments are typical of Kingsley's text, which also frequently mentions eminent Victorian scientists such as Huxley, Faraday, and Darwin, and takes the reader on a journey to encounter evolutionary theory, biology, palaeontology, and other popular Victorian scientific disciplines (see Figure 29), demonstrating that even literary fairies must bow before "the great fairy Science, who is likely to be Queen of all the fairies for many a year to come" (Kingsley: 55).

Our world has changed rapidly since the 19th century; science has advanced swiftly to now include digital and computer technologies, and there are fewer natural spaces for fairies to live in. Yet even today, we see the connection between science and our 21st-century fairies in many of the same ways that the Victorians enjoyed. For example, Buckley's tradition of employing fairies for science education is given a 21st-century twist in the 2019 picture book

Figure 29. Scientists Richard Owen and Thomas Huxley inspect a water baby. Illustration by Linley Sambourne (1885). Image in the public domain.

Fairy Science by Ashley Spires. In this humorous illustrated book, young readers are introduced to Esther, the only fairy in Pixieville who does not believe in magic, preferring instead the rationalisations of science. Esther, we are told, "prefers facts, data and hard evidence to wishing on stars", and the accompanying illustration depicts Esther sitting in her rather traditional fairy hollow in the base of a tree surrounded by her very *not*-traditional scientific equipment, models, and charts, carefully measuring liquid into test tubes. In a playful reversal of the marginalisation of fairy belief in our scientific world, Esther's belief in science places her at odds with her fellow fairies. For example, much to her teacher's frustration, Esther challenges the accepted story of fairies originating from a rainbow by proposing instead a theory of species evolution in response to environment and diet. Where rainbows represent "a path to hidden gold" for the other fairies, Esther sees only "light and water colliding" and offers a neat explanation of dispersion. She similarly explains phenomena such as condensation and erosion, but her attempts to educate the other fairies are ignored until her scientific method proves the only solution to a dying tree, in contrast to the ineffective fairy magic (Esther deduces through research and experimentation that it requires more sunlight). The book concludes with a step-by-step guide for readers to use "the scientific method" and conduct their own experiment to grow a plant from seed. While Buckley employs fairies purely as a metaphor for natural forces, by the 21st century, Spires can be more playful, going a step further than Kingsley's ironic exhortation that *The Water Babies* "is a fairytale", and his readers "are not to believe one word of it, even if it is true" (Kingsley 1994) to allow her fairies to exist so that they might paradoxically teach the scientific method through debunking magic.

Beyond educative books, even the fairies of popular televisual media bear the legacy of the fairies' link with science. One of the most notable examples is *Winx Club*, an animated series centred on a group of fairies attending magic school. One of the most popular fairy-focused forms in recent decades, it spans multiple television series, remakes, and adaptations (including Netflix's recent live-action *Fate: the Winx Saga*). In this world, and particularly in the original animated series, many of the fairies' powers are rooted in invisible forces akin to those identified as the fairies of science by Arabella Buckley. For example, writing of the sudden explosion of plant life after winter, Buckley attributes the "delicate green leaves, [...] nodding bluebells, and pale-yellow primroses, as if a fairy had touched the ground and covered it with fresh young life" to the fairies of science: "the fairy 'Life'", responsible for the growing vegetation, and "the fairy sunbeams with their invisible influence kissing the tiny shoots and warming them into vigour and activity" (151). These magical forces of nature and sunlight are embodied in *Winx Club* by the fairies Flora and Stella. Flora, the fairy of nature, has magical powers that draw on the power of life itself, most typically growing and manipulating plant matter such as flowers and vines (used to ensnare her opponents), but also bestowing life upon inanimate objects. She is also in tune with the natural world, able to hear its voice and emotionally feel its pain upon destruction. The name Flora, of course, evokes the scientific term for plant life; her knowledge of the scientific field of botany is extensive, and she is often depicted experimenting with magically growing different plants. Similarly, Stella is the fairy of the sun, stars, and moon, and her magic takes the forms of light and heat, most typically resembling the glowing yellow light of a sunbeam when cast. She can form this magical sun energy into blasts of light, solar heat, glowing light, and shields of light, and she also explicitly draws her energy from the sun, becoming weakened in darkness.

While Flora and Stella, and even Musa, the fairy of Music, who draws on soundwaves, resemble the invisible forces identified as fairies of science by Buckley, the *Winx Club* series' most explicit link between fairies and science is Tecna, the fairy of technology, who wields "digital powers" and can occasionally

The Fairy-Land of Science (Arabella Buckley, 1878)

Figure 30. Tecna, the Fairy of Technology, uses her digital powers to shield fellow fairies Stella, Flora, and Musa. *Winx Club*, Season 1, Episode 2, created by Ignio Straffi (Nickelodeon 2004).

manipulate electricity (see Figure 30). When cast, her magic takes the form of bright green light, evoking the lurid green text of early computer screens, and forming zigzag shapes that evoke early electronic displays like the cathode-ray tube, or geometric and network or grid shapes that evoke the contemporary mathematical and computing realm of the internet and digital technology. One of the first spells we see Tecna cast, for instance, is a defensive spell that imprisons her enemy within a green ball resembling the World Wide Web symbol, and she similarly conjures a shield of translucent green light called a "Firewall" in a reference to the network security device that blocks malicious traffic to a computer. Tecna can connect her mind to digital and computer networks, access a magical information database, and is naturally adept at manipulating machines and technological devices (see Figure 31).

More than this, Tecna's associations with science and technology extend to her personality. With a highly logical, rational, and calculating mind, she is less adept at navigating her emotions, thereby tapping into a long history of science fiction literature and media that associates machines and digital forms with the nonhuman, robotic and artificial (and indeed, in one episode, Tecna

Figure 31. Tecna, the Fairy of Technology wields her digital powers. Image credit: Winx Club, Season 1, Episode 2, created by Ignio Straffi (Nickelodeon 2004).

is magically transformed into a robot, albeit with disastrous consequences). Importantly, Tecna's digital powers draw on the same ideas of an all-pervasive yet invisible force as the other fairies' more nature-based powers involving the forces of life, light, and soundwaves. While the green colour of her magic evokes the digital realm, it also links her to, and, by extension, places this digital realm alongside, the natural world: Flora's nature-based magic is also green, participating in a history of using the colour to link fairies with the natural world that, as Diane Purkiss notes, dates to medieval times (2007: 73). Alongside sunlight, sound and nature, the digital network of internet communication has become such a ubiquitous force in the 21st century that if Buckley were still writing her science education books today, she would no doubt have included a Tecna-like fairy to explain this phenomenon.

The connection between fairies and science has a long history, and it is one that often plays upon the invisible nature shared by both fairies (as hidden, invisible, or imagined beings) and the scientific forces that make up the natural (or technological) world. If science can acknowledge that invisible forces like gravity, electricity, and sound exist in our world, influencing matter as if by magic, then why cannot fairies also exist, perhaps flitting just beyond the reach of the telescope or microscope? Such is the hopeful possibility conjured when we link fairies with the invisible forces of science.

Part VII

Environmental and Ecological

J. S. Mackley

"Sir Orfeo" (Anon., Late 13th/Early 14th Century)

Medieval Literary Representations of Fairies and the Green Children

Fairies are an integral part of Britain's folklore, particularly in the Celtic nations. However, early descriptions differ from the representations of mischievous but benign angel-like entities, who flutter about granting wishes, and with fairyland as a welcoming, ethereal place akin to childish imaginings of heaven. This chapter considers the introduction of the word "fairy" into the English language, along with an encounter with two fairy-like children as chronicled by William of Newburgh in his *History of English Affairs*. In addition, it will discuss two early poems describing Faerie, "Sir Orfeo", dating from around the beginning of the 14th century, and "Thomas of Erceldoune", dating from the end of the 14th century.

    Medieval representations of fairies are different to those of the post-Victorian era. Leslie Ellen Jones argues the people of the Middle Ages considered fairies as belonging to a "secular component" of the supernatural realms, whereas angels and demons belonged to the "religious sphere" (Jones 2002: 128). Over the centuries, the image of fairies became coated with a veneer of Victorian morality. Nicola Bown notes ghosts and spirits were manifestations of the dead, but fairies "had nothing of the ghastly power of the dead awakened" (Bown 2001: 2). Authors such as Rudyard Kipling and C. S. Lewis attempted to slow the transformation from the earlier depictions of fairies, observing how late Victorian representations were "tarnished by bad pantomime" (Lewis 1964: 123). For Puck in Kipling's *Puck of Pook's Hill*, the People of the Hills represent the land's true nature spirits and Puck grumbles they "don't care to be confused with that painty-winged, wand-waving, sugar-and-shake-your-head set of impostors" (Kipling 1906: 12).

Etymologically, the noun "fairy" came into English around 1330, deriving from Old French *faerie* meaning "enchanted" and referring to the "fairy realm". A more distant linguistic root is the Latin *fata*, also the root of *fate*, and refers to the *moirai* from Greek mythology, who measured and cut the lengths of man's existence. Before the 14th century the word *aelf* denoted supernatural entities which were venerated in pre-Christian Saxon homes as a kind of ancestor deity. These were known as *hus-aelfs*, but were not always benign: *Beowulf* links elves (*ylfe*) with giants and monsters "eotenas ond ylfe ond orcneas", (l. 111), and the Low Middle German word *alf* referred to an incubus, a male sexual demon. The *South English Legendary* describes how the angels who sided neither with God nor Lucifer were "demoted" to dwell on earth and C. S. Lewis argues this act was the degeneration of angels into fairies (Lewis 1964: 136). This corresponds with the apocryphal Old Testament story found in 1 Enoch where "Watcher Angels", called Grigori, come down to earth and impregnate the women; their offspring are malign giants.

Christianity was uncomfortable with fairies, and so they became aligned with Neutral angels and Watcher Angels and were presented to worshippers as dangerous to the immortal soul. As the *ylfe* were already considered sexual demons, this was then transported to medieval romance. Female fairies and their ilk became enchantresses and temptresses, as depicted by Morgana le Fay, for example. Geoffrey Chaucer's *Wife of Bath's Tale* describes a land once "fulfild of fayerye" and the elf queen and her entourage danced in fields, but the narrator laments "lymytours" [licensed beggars] and "othere hooly freres" [friars] now occupy the places where the fairies once were. This is symbolic of the Christian suppression of pagan religions in Britain. As Chaucer demonstrates, *corrupt* religious officials drove away the fairies (ll.857–61, 866).

At the time Chaucer was writing, green was considered an unlucky colour and was associated with otherworldly entities (Brewer 1997: 183). In *The Friar's Tale*, the Summoner meets the devil wearing "a courtepy [short jacket] of grene" (l. 1382), while, in *The Golden Targe* (written c.1508) William Dunbar describes "Pluto the elrich incubus in his coat of grene" (14. ll. 8–9). The devil is represented as green skinned (see Figure 32) in Michael Pacher's altarpiece depicting St Wolfgang and the Devil (1471–1475). The otherworldly

Figure 32. Michael Pacher: St Wolfgang and the Devil (c. 1475).
Image in the public domain.

nature of the colour green is illustrated by the story of the green children of Woolpit, Suffolk, chronicled by William of Newburgh in the *History of English Affairs* (c.1189). William never uses the word "fairy": he writes *de viridibus pueris*, literally, "concerning the green children". Another chronicler, Ralph of Coggeshall claims the source for this episode in his *Chronicle of the English* (early 1220s) was Richard de Calne of Wykes, who took the green children into his manor during the reign of Stephen de Blois (reigned 1135–1154).

The legend describes how two children with green skin appeared near one of the wolf pits from which the village takes its name. They spoke no English, nor did they eat. The boy soon died, but the girl was eventually able to eat human food, and her skin lost its green hue. When she learned English, she explained she had come from an underground country with "neither sun nor moon but a soft light like twilight" (Briggs 2002: 9; cf. Larrington 2019: 209–11). The children had been herding their flocks when they followed the sound of distant bells and went through a cavern until they reached our world and were dazzled by the sunlight. She also claims the inhabitants of her land were Christians. Still, the land is separated from our world by a wide river (Briggs 2002: 9). In medieval legend, the "land across the water" was indicative of a Paradise towards which men could travel and see for themselves but were prevented from exploring too far because it was beyond their human understanding (Clark 2024: 313; Harte 2024: 41–2).

Derek Brewer posits there may be a rational explanation for the green children, suggesting the lost children were suffering from chlorosis or "green sickness" (Brewer 1997: 182). However, the story of these children is a moral of redemption and demonstrates the power of Christianity over pagan religions. It is also contrary to the fabled fairy practice of abducting human children and leaving changelings in their place. It does not emphasise the terrible outcomes of eating otherworldly food (after all, the green children must think of humanity as "otherworldly"). The children receive baptism, and the girl flourishes in her new world, thus demonstrating even supernatural creatures can achieve salvation.

Sir Orfeo

Just as the story of the Green Children shows how they found their way into our world, the story of "Sir Orfeo" dating from the late 13th century retells the classical story of Orpheus and Eurydice entering the classical underworld as told in Virgil's *Georgics* and Ovid's *Metamorphoses*. In these versions, Eurydice is abducted and Orpheus travels into the nether realms and through his "irresistible" music, he persuades Pluto to release Eurydice. This is conditional on Orpheus leaving the Underworld with Eurydice following behind him, and him not turning back. Unfortunately, he does turn, and Eurydice is whisked back to remain in the Underworld forever.

The medieval narrative of "Sir Orfeo" differs from the classical version. The story has been transposed to England. Orfeo is a good king and is renowned for his skill at harping ("A better harpour in no plas" l. 32). However, when Heurodis, his queen, awakes under an "ympe-tre" (a "grafted tree" l. 70), she describes a nightmare, telling her husband the king of Faerie will take her to his realm, either willingly or by force. Here, "faerie" refers interchangeably to the supernatural folk, the land where they dwell, and the enchantments they cast.

Of course, Heurodis does not wish to leave her husband, but she describes Faerie as beautiful and bountiful. She says the king of Faerie "made me with him ride" (l. 155) and showed her Faerie as a palace with castles and towers along with an abundance of natural beauty (ll. 161–2). However, Heurodis must accept her fate: "And yif thou makest ous y-let,/ Whar thou be, thou worst y-fet,/ And tortore thine limes al" (And if you hinder us, wherever you are, you will be brought/ and your limbs will all be torn apart [169–71]).

The symbolism of the "ympe-tre" under which Heurodis slept has been much debated (see, for example, Curtis 2008: 143; White 2013: 6). It is normally translated as a "grafted" tree. However, it is unclear why falling asleep under such a tree is a transgression with terrible consequences. It may have been a meeting place of humanity and nature (grafting representing humanity's attempt to control nature) and a liminal point between the natural and the supernatural realms. Alternatively, eating the fruit was prohibited, as was the case with Eve in the Paradise of Eden; or perhaps it is assumed Heurodis ate the fruit at this threshold and, like the green children, trapped herself in the fairy realms. The motif of the tree being a border between the human realm and Faerie, as well as the prohibition of eating fairy food, are also themes found in "Thomas of Erceldoune".

Despite an army of well-armed knights, Orfeo cannot prevent Heurodis's abduction, she is taken to Faerie, but no one sees how. Orfeo abandons his kingdom and responsibilities for "ten yere and more" (l. 264), leaving a steward in charge while he wanders through the forest, taking solace in playing his harp. After a decade, he sees "the king o fairy with his rout (company)" and Heurodis riding among ladies on horseback, and he follows them back to their faerie realm.

Orfeo follows the entourage to the otherworld "at a roche" (l. 347), he does not need to perform a ritual, for example dancing three times *widershins* around a fairy mound (Jacobs 1891: 191; Menefee 1985: 8). Once inside Faerie, he travels for three miles until he reaches a bright country with the wonderful castle with a hundred towers of which Heurodis dreamt. This imagery is a common trope in medieval literature, borrowing descriptions of the New Jerusalem in Revelation 21; indeed, Orfeo declares it is "the proude court of Paradis" (l. 376). However, once inside the castle gates, Orfeo witnesses a different scene: the inhabitants are suspended in the moment they were taken from our world. Some are without heads, or arms, others with bodily wounds; some are mad, others strangled, some drowned, and others burned by fire, and some taken as though by enchantment (ll. 391–400). The land of Faerie is represented as the classical underworld to the extent that its rulers are Pluto and Proserpina. For the Medieval Christian, the image of dismemberment after death needed to be overcome, because Christian doctrine promised the body would rise "incorruptible", which meant it would rise "whole" at the time of the Resurrection, and not dismembered as depicted in "Sir Orfeo" (Bynum 1990: 52)

The Lord and Lady of the Underworld are pleased to welcome him, and after Orfeo beguiles them with his harping, Pluto allows him to leave with whatever he chooses. When Orfeo announces he wishes to take Heurodis, Pluto objects they would not make a good couple, but consents and wishes "of hir ichil thatow be blithe" (I wish that you would be happy, l. 471)

The audience may have anticipated the dark ending of the classical versions; however, in the story of Orpheus and Eurydice (along with its resonances with the Biblical story of Lot and his wife fleeing the destruction of Sodom in Genesis 19) there is the condition of not looking back. This can be seen as the victim wavering when being given a command or breaking a taboo ("do not look behind you"), or secretly longing to return to the sinful ways. Orfeo leads Heurodis to Winchester without condition or incident. Thus, rather than highlighting the continuing oppression against humanity, "Sir Orfeo" illustrates the turning of the Wheel of Fortune – the narrative begins where Orfeo is successful and happy. The Wheel turns, Heurodis is abducted, and Orfeo abandons his kingdom, but Heurodis is restored to him. In the final section, Orfeo discovers that the steward in charge of the kingdom has remained faithful to him, and thus, Orfeo is restored as king. "Sir Orfeo", therefore, is the retelling of both the classical and Biblical stories without the conditions affixed on both to return, and so the story has a happy ending. However, the Wheel of Fortune remains turning, and Orfeo realises he must die one day, but the good steward and his family will rule in Orfeo's place.

Thomas of Ercildoune

In contrast to Heurodis forcibly being taken to Faerie, the Scottish poem "Thomas of Ercedoune" (c.1400), describes how Thomas (known as "the Rhymer" or "the prophet") is taken to Faerie. Like Heurodis, Thomas lays beneath a "semely tree" (Murray 1875, l. 34) which is believed to be the Eildon Tree, which stood at the foot of one of the Eildon hills near Melrose in the Scottish Borders. Thomas sees a "lady gay" rising towards him and calls her the "Quene of heuyn" (l. 88) but she quickly explains that she is "of a nothere contre" (l. 93). Thomas requests to lie with her even though she implores "I pray the Thomas, lett me be!" (ll. 102, 128.)

The lady leads him from the Eildon hill into a land that is abundant with fruit and beautiful birds, but he is warned "þe deuele wole the ateynte" (the devil will come for you) if he eats the food there (l. 188). So, as with the description in "Sir Orfeo", this realm is filled with dangers, and this motif links with the tradition that eating fairy food traps the mortal in their realm. She shows him pathways leading to heaven and to hell, as well as her own beautiful castle, although she fears the repercussions of her lord, the king, if he "wyst þou layst me bye" (knew you had lain with me, l. 224). Although Thomas believes he has been in this land for three days, he is informed it has actually been three years. The different passing of time in Faerie compared to the earthly realm is another common motif in fairy literature. However, Thomas is told he cannot stay any longer as the devil will come into the land and take his "fe" (a tithe or sacrifice). The lady is convinced Thomas will be a human sacrifice: "he wyll haue þe" (l. 292). Thomas agrees to leave by means of the Eildon tree but asks for "sum tokyne" to remind him "that ever I saw the" (l. 312). She offers him the choice "harp or carp"– to sing or to speak; this again links with Orfeo's skill at minstrelsy. Thomas chooses the latter and the lady tells him of future events to occur in Scotland:

> "Who shalbe kyng, and who shall be none,
> And where any battell done shall be,
> Who shall be slaye, who shall be Tane [taken],
> And who shall wyne the north Contre". (ll. 345–8)

Thomas returns to his own realm and the lady's gift becomes the foundation of his prophecies.

Thomas's first instinct was to sleep with the lady, but this text demonstrates that female fairies are not always seductive demons. They are capable of love with mortal men and are prepared to offer them great gifts when their love cannot continue. This contrasts with later texts such as "Goblin Market" and "La belle dame sans merci", where the fairy enchants and then ensnares the knight. The action in "Thomas of Erceldoune" takes place at a liminal point both between heaven and hell, as well as her Lord's castle and the human realms, although, as the Lady observes, the devil is still a threat in these lands.

In conclusion, the term "fairy" appeared in English in the 14th century, although the folkloric "wee folk" had been around much longer. However, fairies belonged to the past. The descriptions of the fairy otherworld in "Sir Orfeo" and "Thomas of Erceldoune" are set in a legendary past; the descriptions of the green children are comparatively contemporary to the chroniclers that recorded them. Fairies were associated with the colour green, and so were early representations of the Devil. Likewise, the Realm of Faerie to which both Orfeo and Thomas travel are linked with the Classical underworld or a purgatorial, liminal place where the threat of the devil was a constant undercurrent. However, as Bown notes: "Fairies always belong to yesterday because today's society is corrupt, sophisticated, urbane and disenchanted" while fairies belong to a time of innocence (Bown 2001: 163). This loss of innocence was also symbolic of humanity's expulsion from the Paradise of Eden. However, because fairies belonged in the past, there was no need to fear them today: the descriptions could employ classical imagery and are exemplars of the power of Christianity over pagan religions. That said, the fairies' characteristics would change over time, and they would soon become the mischievous sprites as the stories about them developed.

Morgan Daimler

The Call (Peader O'Guilin, 2016)

The Irish Sídhe Through a Folkloresque Lens

Urban fantasy as a genre has existed since the 1980s, popularising and giving new life to a range of beings previously found only in localised folklore and anecdotal accounts. Among the beings that have become staples across urban fantasy are the Irish sídhe, who are often conflated with or understood to be fairies but who occupy a very different space in literature and folk belief than the popular idea of fairies shaped by the Victorian era. Rather than twee-winged children the sídhe in fiction are depicted as human-like magical beings who interact with human protagonists much as other human characters would, but often with the addition of folkloric flourishes, such as a weakness to iron. These fictional sídhe are read and absorbed back into belief by people without a strong grounding in existing folklore, creating a tension between the two often antithetical views. This can be illustrated by examining the folklore and a novel based on it, *The Call*, compared with the folkloresque and a novel grounded in that, *Born To Run*.

The Sídhe in Folklore

Across Irish folklore, the sídhe, or more appropriately the aos sídhe (people of the fairy hills), are portrayed as mercurial and powerful beings who are not only intrinsically tied to humanity but also, in many instances, in opposition to humans they encounter. Katherine Briggs includes one account by a man from Donegal who said of the sídhe: "[they] are not earthly people; they are people with a nature of their own" (Briggs 1967: 176). Similarly, Linda-May Ballard in her article "Fairies and the Supernatural in the Reachrai" succinctly describes these beings in Irish belief as "... a parallel race of beings, capable of helping humans, but capricious and best avoided" (Ballard 1991: 48). These were beings who made their homes in fairy trees and fairy hills or forts, who were literally and figuratively embedded in the Irish landscape and who represented powers both intimately familiar and impossibly foreign (see Figure 33). The sídhe existed alongside humanity for both good and ill, sometimes reaching out for assistance to their human neighbours – often seeking the loan of something the sídhe lacked – but sometimes causing harm in the form of illness, madness, death, or the theft of living humans (Ó'Giolláin 1991: 201–2). The theft of humans, both children and adults, was a particular activity of the sídhe which was noted in folklore across Ireland as well as in fairies across western Europe more generally; those taken were usually considered especially beautiful, well mannered, or otherwise exceptional which made them targets for the sídhe who wanted to include that human among their own numbers.[1] Various anecdotal accounts include mentions of humans thought to be dead or taken by the sídhe who were later seen among the fairy throng, and many versions of the ubiquitous 'Borrowed Midwife' story hinge on the midwife being called to attend a labouring woman who she recognises as a local girl thought to have died or been stolen. While the sídhe often feature in stories across folklore and into modern anecdotal accounts they are always warned against as a possible danger.

---

[1] There are some less pleasant variant beliefs about why humans were stolen and found in Scotland; however, the focus here is on the Irish beliefs specifically.

Figure 33. "The Banshee" by Henry Justice Ford, 1902. Image in public domain.

The Folkloresque

Folkloresque, as defined by Michael Dylan Foster in the introduction to the book *The Folkoresque: Reframing Folklore in a Popular Culture World*, is material which draws on folklore as a source to reference wider cultural material, but which is itself not folklore and which exists outside the usual informality of folklore (Foster 2016: 1). This material exists in an interesting limbo between the tradition of folklore, on which it is loosely based, and the vast sea of fiction in which it is placed, belonging wholly to neither. The sídhe in folkloresque material can run a gamut from what is found in O'Guilin's work, which adheres more closely to traditional material and may stand as an example of folklore within fiction, to works which show them as caricatures of the beings of folklore such as we find in Lackey's novels, so heavily rewritten that they are connected to the folklore only by the name they are called.

Finding Folklore in Fiction: O'Guilin's *The Call*

Peadar O'Guilin is an Irish author who has written a two-book series which incorporates the sídhe as key antagonists. The premise of the book *The Call* is that thousands of years ago humans drove the sídhe from earth, and they have never forgiven that loss and are working to return. His sídhe, while clearly anchored in the fictional reality of his alternate earth Ireland, are largely described in line with older folklore. They are indescribably beautiful, compared to royalty and fashion models, and also incomparably cruel, stealing human children away to hunt them for vengeful sport. Some of these children survive and return to Ireland, but others are killed or horrifically disfigured, treated like malleable clay to entertain the sídhe who are trapped in their own world but yearn for a return to the human one. One girl has her nose and mouth erased as if she never had them and dies; a boy survives but is grotesquely distorted, turned halfway into a giant before being pulled back to earth; one girl is returned alive with a permanent, fist-sized hole through her torso (O'Guilin 2016: 261, 223, 227). This interaction encapsulates O'Guilin's sídhe: "The newcomer is the most handsome man she has ever seen. His face is kind and full of humor as he kicks her in the side hard enough to knock her over onto her stomach. He laughs." (O'Guilin, 2016, 284). As with the sídhe of folklore they do make deals with humans, giving a person something that is desired in exchange for something that the sídhe want or need. The sídhe of *The Call* are beings of another world who seek to regain a place within the human world and yet they are utterly foreign to earth, beings that are not only magical and powerful but also capricious and vicious (see Figure 34). They are then, like the sídhe in folklore, feared and

The Call (Peader O'Guilin, 2016)

Figure 34. Black and white illustration by Arthur Rackham for James Stephen's *Irish Fairy Tales* (1920). Image in the public domain.

respected by the humans who encounter them and who have little choice but to play a game they cannot win and do not fully understand.

The Sídhe in Fiction: Mercedes Lackey's *Born To Run*

In contrast to O'Guilin's folkloric figures we find the Irish sídhe in other works appearing through a stronger folkloresque lens, particularly in urban fantasy, in ways that divorce them from their original folklore and create a new vision of these beings based on the author's creativity and plot needs. An example of this is the Serrated Edge series by American author Mercedes Lackey which begins with the novel *Born to Run* published in 1992. Lackey's sídhe are clearly established as Irish in origin and magical in nature, with some folklore being used to fill in details, but are largely the author's own creations. For example, while folklore does establish that many fairies including the Irish sídhe are averse to iron, a detail that Lackey also incorporates into her work,[2] the sídhe of *Born to Run* are also negatively affected by caffeine which acts on them like a narcotic. And while the sídhe in folklore are described as looking very much like humans, able to pass among humans when they choose to, Lackey's sídhe are described typically in this fashion: "the man had bright, emerald green eyes; eyes that looked just like a cat's. And long, pointed ears" (Lackey 1992: 40). With this description Lackey inserts two blunt physical cues for readers – and her human protagonists – to quickly identify characters that are sídhe, sidestepping the problem presented in folklore of describing what is often a physically imperceptible difference in nature.

Lackey's sídhe are beings who exist in close connection to humanity by choice; beings from another world which they call "underhill" who lack creativity and rely on humans to inspire them and keep them from being lost in a state of mindlessness (Lackey 1992: 71). They are ruled by a high king, Oberon, and while the Irish word sídhe is repeatedly used to describe them Lackey has them existing across western Europe and synonymously refers to them as elves. Perhaps most striking of Lackey's adaptations are the application of the Scottish Seelie and Unseelie courts[3] to the Irish sídhe and her explanation for why the sídhe steal human children, a concept that is well established in folklore. In Lackey's series, there are two distinct groups of the sídhe, under the Scottish terms, with the Seelie being generally benevolent and helpful to humans and the Unseelie dedicated to anything and everything that causes humans harm (Lackey 1992: 49–50). As part of the wider benevolence of the Seelie, Lackey explains through a bit of expositional dialogue that the stories of the sídhe stealing children are rooted in their need to rescue human children from abusive situations, driven by their own low birth rate, and protective instincts (Lackey 1992: 45–6).

Ultimately Lackey's sídhe are an entirely new creation that is only barely clothed in the folklore of the old, using terms taken from Irish and a few highlights of popular belief which are rewritten to fit more effectively into the world Lackey is creating.

2   At least in theory. The author establishes this as a key weakness of these beings in comparison to humans however did not seem to be aware of the range of human products that contain iron beyond the most obvious.
3   Spelled seleighe and unseleighe in Lackey's work but standardised to Scots here.

The Call (Peader O'Guilin, 2016)

Divergent Views

What exists then across modern fiction is two divergent views of the Irish sídhe, one which adheres more closely to the understanding of them in folklore and one which embraces a folkloresque understanding and sees the sídhe as malleable to the author's needs. Material which follows the first approach, as with Ó'Guilin's work, depict the sídhe as inhuman and inhumane and often acting inexplicably or in contrast to established human mores. In contrast, the folkloresque material, as seen in Lackey's work, anthropomorphises the sídhe and portrays them as something similar to magical humans, using the folklore only where it is advantageous to the author and discarding or modifying it elsewhere.

The second view of the Irish sídhe as found across popular culture and particularly in urban fantasy has created a new understanding of these beings divorced from Irish folk belief and framed instead entirely within the fictional narrative. The sídhe become not the enigmatic, dangerous beings of the Irish cultural milieu, but are transformed into thoroughly modern anthropomorphised figures who are defined and limited by their interactions with human protagonists. These newly defined sídhe are then taken from fiction into belief in a process similar to the one discussed by Mullis in the article "Cryptofiction! Science fiction and the rise of cryptozoology" wherein ideas put forth in fiction are adopted into real-world belief by those seeking explanations for phenomena around them (Mullis 2019: 247–8). In this process, readers of fiction featuring the sídhe incorporate the beliefs about these beings as outlined in the novels into their own personal beliefs, using fiction as a foundation to understand particular phenomena through the specific lens of fairies. Then, in turn, new authors pull on these folkloresque depictions rather than the root folklore and perpetuate the cycle, giving us popular ideas of fairies with wings or sídhe with pointed ears which are not found in older folklore but began in theatre, art, and fiction (Daimler 2020: 267–71; Young 2019: 266–-7). The folkloresque, if it is accepted widely enough and for long enough, becomes the folklore.

The Question of Appropriation

An unavoidable question raised by this process of removing the sídhe from their cultural context, redefining them, using the new versions to populate fiction, and then having that new version find widespread acceptance is whether what is being done is appropriative of Irish folk beliefs. There is an argument that can be made that this popularity of the sídhe in today's media represents cultural diffusion rather than appropriation, but that view often overlooks the history of colonialism and exoticism that has long affected the Irish culture and which is a key factor in the use of the sídhe in fiction. Cultural diffusion "takes place in piecemeal transactions between relative equals without coercion being the dominant motivation", while appropriation is a situation where "the powerful group takes aspects of the culture of the subordinated group, making them its own" (Baird-Jackson 2021: 87–8). Appropriation is also characterised by anger or unhappiness in the source culture over this material being taken, and a marked indifference or lack of awareness in the culture doing the taking (Baird-Jackson 2021: 88). The most popular authors who include the sídhe in their urban fantasy are generally not Irish themselves, most often American, and as in the example of Lackey's sídhe usually largely redefine who and what the sídhe are to work with their own plot needs. Orla Ní Dhúill discusses this problem in depth in her article "Do Fantasy Writers Think Irish Is Discount Elvish?" which points out the use of both the Irish language and Irish folklore as a shortcut to signal the fantastic within a story, where the language is used as a fantasy language and the folklore, heavily modified, to quickly signal to the consumer that something magical is afoot, even when it bears little resemblance beyond the terms to the actual folklore. Through this the word sídhe itself becomes a cue that something fantastic is involved, implying the sídhe's innate magical nature and foreignness from humanity (see Figure 35). This taking by those outside the culture because of the perceived exoticness and repackaging without understanding into a new product for the other culture to consume fits the description of appropriation and often appears intrinsically within folkloresque material drawing on Irish sources.

Figure 35. John Duncan's "Riders of the Sidhe" (1911). Image in the public domain.

Conclusion

Fiction will always, and arguably has always, drawn on folklore and folk belief as fuel to inspire creators just as folklore has always existed as a way for groups to share beliefs and practices. This creates perhaps, at best, a kind of cultural or folkloric diglossia where the older folkloric beliefs and the new popular culture beliefs connected to the word sídhe exist side by side as two different cultural forms acting in effect like two different languages so that how a person interprets the word sídhe will depend entirely on whether they understand through a folklore or folkloresque lens. This phenomenon is not a new one, having been seen in various places where there is tension between folk belief and institutionalised beliefs, where the folk belief represented one cultural force and the institution another (Ó'Goilláin 1991: 209). However, the current iteration being played out with the sídhe between folklore and fiction with the mediating factors of social media, publishing, and the internet reflects a sharper contrast and more extreme variance than has perhaps been seen previously, as the older cultural folklore is overwhelmed by an outside flood of folkloresque fiction. This tension is further highlighted by works which adhere more closely to folklore and those who use it as the barest guide for creative interpretations. O'Guilin shows that it is possible to include the more folkloric sídhe in a work of fiction without losing the overall tone of the folklore while Lackey is an example of the problems endemic in moving too far from the material and ultimately redefining the entire concept instead of including it as it exists within the source culture. Both contribute to the wider Western cultural understanding of the sídhe outside of Irish folklore; it remains to be seen which understanding of the sídhe will ultimately dominate the folklore of generations to come.

Sophia Lange

Magic: The Gathering (Richard Garfield, 1993–Present)

Dreams, Memories, and Matriarchy

Often considered the first of its kind, the collectable trading-card game *Magic: The Gathering* (*MTG*) has amassed an estimated 35 million players worldwide since its establishment in the early 1990s. The battle game published by Wizards of the Coast and designed by Richard Garfield centres on a duel between opponents that "simulates a magical conflict between two wizards" (Golub and Peterson 2016: 333). In this conflict, spells (read: cards) are summoned from the players' card decks to influence the course of the game, ranging from different creatures of various strengths and with various abilities, instantly damaging or healing spells, and protective or destructive enchantments, to so-called board wipes which destroy all cards in play. Given this wide range of possible courses a single game may take, players must create their decks carefully by selecting cards with supporting effects or thematic alignments. As pointed out by Dodge and Crutcher, "[d]eck construction is strategic in a concrete way, but it is also an act of creation rich in background knowledge, meaning, various literary practices, imagination, and personal flair" (2018: 170). One of the game's structuralising components is the different species that may guide the creation of a player's card deck, since the species are interconnected through larger narratives contrived in *MTG*'s accompanying novels and spread via Wizards of the Coast's media outlets.

In 2007, Wizards of the Coast published the 43rd expansion pack to *MTG*, based on three novels created by Cory J. Herndon and Scott McGough, namely *Lorwyn* (2007), *Morningtide*, and *Eventide* (both 2008). This expansion was inspired by Celtic mythology and introduced more than 500 new cards, with not one human creature card amongst them. Over the past decades, *MTG*'s artists have regularly "adapted cultural myths and tropes into individual card sets, or wide card settings" (Zanescu 2022: 1), so it came as no surprise to see the so-called Lorwyn/Shadowmoor expansion pack feature a decided nod to the lore of the British Isles. However, as Zanescu rightfully points out, "[t]he issue that arises when depicting the habitus of an entire culture or epoch in a few hundred cards is predictable enough: stereotyping" (Zanescu 2022: 1). Even though *MTG* has recently received scholarly attention, for example, regarding its potential as an educational tool for second language acquisition (see Dodge and Crutcher 2018: 171), an analysis of the game's twist on mythological lore, its resulting character formations, as well as its potential impacts on the realities of *MTG*'s players, however, has largely been omitted. While it would exceed the scope of this contribution to relay the multilayered representations of cultural stereotypes at work in the particular expansion pack in question, I will focus on one distinct creature type which holds special significance for the conceptualisation of *MTG*'s narrative universe.

The fairies, with Queen Oona as their matriarch, her name a decided nod to Una, the last High Queen of the *aos sí* in Irish folklore, are one of the key species of *MTG* and have become a favourite of many (see Gravelle 2019). As a strong and beautiful matriarch, Oona can be aligned with other fictional fairy queens such as Edmund Spenser's *Faerie Queene* (1590), William Shakespeare's Titiana from his *A Midsummer Night's Dream* (1595/96), or his Queen Mab, famously featured in *Romeo and Juliet* (1597). While Shakespeare had promoted the concept of a fairy's miniscule size, causing it to be "reduced to a miniature, winged denizen of flower buds and nursery illustrations" by the 19th century (Forsberg 2015: 639), *MTG*'s Queen Oona is described as a creature of enormous size with the head and torso of a beautiful female and her lower body being an assemblage of lush vegetation (see Herndon 2007: 285–7). As a flower–fae hybrid, *MTG*'s matriarch is thus inherently connected to her natural surroundings, which echoes classical Victorian depictions of fairies, representing "a means of imaginatively reconceptualising the natural world as a place of minute wonders" (Forsberg 2015: 640). In contrast to this idealised understanding of fairies as delicate and wondrous creatures, Oona is a powerful monarch who rules over her subjects with an iron fist and seeks to expand her influence to become sole empress of Lorwyn, the plane on which the aforementioned novels by Herndon and McGough are set. Since other fairies in *MTG* are indeed of miniature size and often confused with insects, Oona has resolved to more subtle means of extending her hold on Lorwyn than mere force. According to *MTG*'s lore, relayed in detail in both the novels and on the game cards themselves, Oona's magical power is "nourished by secrets and pollinated by stolen dreams,[1] which are harvested by her subjects to whom she refers as her children (see Herndon and McGough 2007: 177). Given that *MTG*'s fairies are unable to dream themselves, Oona's children must infiltrate the thoughts and dreams of others to feed their queen, which aligns the matriarch with Queen Mab as the fairies' midwife in Shakespearean folklore, who helps sleepers "birth" their dreams to make them do her bidding.

Unlike most other races on Lorwyn, fairies are thus neither benevolent nor inherently good, but seem to exist outside of general estimations of morality. This further connects them with Irish mythology, which perceives fairies as awe-inspiring but frightening creatures that are immensely protective of the natural spaces they inhabit. In Oona's kingdom-as-ecological-utopia called "Glen Elendra", reminiscent of Tolkien's "Faerie", "that 'Perilous Realm' where anything can happen" (Tatar 2014b: 4), natural wilderness thrives, unobstructed by outside interference. As outlined earlier, neither the novels nor the expansion pack introduce any humans, meaning that *MTG* not only offers a nonhuman idyll in Lorwyn as such, which finds itself in a state of perpetual midsummer, featuring wondrous forests, gentle rivers, and natural harmony, but also presents the pinnacle of paradisiacal nature in Oona's particular sphere. Glen Elendra is "an impossibly lush and verdant grove" in which "[f]lowers, fruit, and fairies" thrive in abundance (Herndon and McGough 2008a: 70), all revolving around a matriarch who both commands and embodies the sphere itself:

> She was taller than a giant, a graceful female head and torso perched atop a mounted thicket of thick, ropy thorn canes. Fat green leaves and stout, sharp spikes covered her

---

[1] This quote is taken from the flavortext of "Oona's Prowler", a faerie rogue creature card in *MTG* (<https://edhrec.com/cards/oonas-prowler, accessed 20 June 2023>).

lower half. Thousands of colorful flowers carpeted her upper body like a fine gown. Clusters of blood-red berries adorned her wrists, neck, and ears like jewelry. Her face was hidden behind a veil of honeysuckle creepers, but her huge blue-green eyes shone clearly through the curtain. A wreath of sharp-leave holly and jasmine flowers crowned her head. (Herndon and McGough 2008a: 70)

Not only does Oona thus personify the environmental utopia of both her sphere and the entire plane of Lorwyn, but it is also relayed how she can therefore extend her reach throughout the plane, allowing her to "manifest anywhere" (Herndon and McGough 2008b: 292). As the embodiment of nature's abundance, Oona seeks to become more powerful so that the entirety of Lorwyn may come to resemble the ecological utopia of her realm, "alive and green and vibrant" (Herndon and McGough 2008a: 71).

Unbeknownst to everyone except the fairies, however, Lorwyn's utopian state is anything but permanent. As pointed out in the novels, the plane is in fact home to two different realities, namely that of Lorwyn and its dark counterpart Shadowmoor, which follow one another in a perpetually occurring 300-year cycle referred to as the "Great Aurora". Each transition to the nightmarish Shadowmoor, a realm of everlasting gloom whose darkness triggers hostility and warfare, causes the plane's inhabitants to lose their memories of the former world should their life cycle coincide with Lorwyn's drastic shift. The only race remaining unaffected by the plane's transition are the fae and their home Glen Elendra, which is spared the detrimental environmental effects of the Great Aurora and thereby casts *MTG*'s fairies, similar to other popcultural representations of these magical beings, as unmoved by the "alterations of time" and thus connected with an "idea of permanence" (Wood 2006: 282).

Over the course of the novels, it is revealed that Oona herself is in fact the cause of the Great Aurora, since she had sought to extend the natural cycle to enlarge her paradisiacal utopia and dominance over the realm, and had collected the dreams and memories of others to gather enough magic to spin a net of protection over her entire people.[2] That Oona represents ecological

---

2   Dreams thus function as a magical commodity on Lorwyn and are metaphorically represented by the cards in a player's deck outside the game's lore.

permanence is also represented in her song, etched in the memories of her children:

> I am the root. You are but the flower.
> Yours is the will, but mine is the power.
> I am the trunk and the enduring boughs,
> Majesty no fleeting bloom can long house.
> Grandeur beyond what one blossom contains.
> While it fades and withers, Oona remains. (Herndon and McGough 2008b: 190)

Despite her being cast as environmental protectress due to her wilful extending of the Lorwyn/Shadowmoor cycle, thus not only increasing her power but also maintaining the lush splendour of Lorwyn and thereby allowing the ancient treefolk to gather "thousand years of lore and knowledge, and many thousands more only half-remembered" (Herndon and McGough 2008a: 58), the fairy matriarch remains a much-contested figure in *MTG*'s lore. Her status as ecological warrior queen seeking to dominate her plane via psychological trickery aligns her with Maria Tatar's reading of postmodern female tricksters as double agents who employ "strategies both subversive and transformative in order to construct their own identities but also to effect social change" (Tatar 2014a: 46). Granted, Oona's aim is superficially concerned with environmental rather than social improvement but considering the effect of environmental alterations on the inhabitants of her plane, her desires are also clearly aligned with socio-political shifts. While she selfishly seeks omnipotence, an extension of Glen Elendra as a blissful sanctuary would also mean a manifestation of Lorwyn's never-ending growth season and a continuation of peaceful harmony amongst the plane's races. The fact that she operates under the radar, since her realm is invisible to non-fairies (see Herndon and McGough 2007: 283), further casts her as an early version of female tricksters, having "carried out her own stealth operation, functioning in furtive ways and covering her tracks to ensure that her powers remain undetected." (Tatar 2014a: 40). Given her well-known presence throughout Lorwyn, it must be acknowledged that even though she casts a veil of shadows over both her realm and her ecological quest, her dominance is felt everywhere, and she therefore becomes representative of a cultural change that has acknowledged female agency and power (see Tatar 2014a: 52).

As much as Oona's powerful hold on Lorwyn inspires others to perceive her as a destructive "[d]ream-spinner, glamer-weaver", who has made the plane her playground and its inhabitants her toys (Herndon and McGough 2008a: 192), it must be noted that the particular Great Aurora relayed in *MTG*'s novels is more devastating than previous shifts. In a terrifying example of contemporary ecohorror, Lorwyn transforms into an anthropocenic wasteland defined by destruction:

> Change crawled south across the landscape like some terrible wind, slow but inexorable, entirely irresistible. Sunlight-dappled meadows and lush green forests of oak gave way to stale, murky bogs and gaunt, haunted willows. Wildflowers curled and sharpened into brambles that were tipped with glistening red. The pure, clean flow of the Wanderwine slowed to a dank, muddy crawl in the south, while up north it raged and frothed through a new series of white-water rapids (Herndon and McGough 2008b: 243).

The effect of this specific Great Aurora is particularly devastating on Lorwyn's ancient treefolk, harbingers of wisdom and memories, who have now "become rotten with cankerous disease and bitterness" (Herndon and McGough 2008a: 9). That Oona had originally sought to prolong the cycle, therefore, would actually speak for her attempt to retain the lush pastoral paradise that is Lorwyn, including its knowledge of environmental harmony personified by the ancient treefolk. One may consequentially read the way in which the Great Aurora's cycle has slipped away from her grasp as a play on contemporary notions of climate change and environmental destruction in the Anthropocene. Various characters in the novels tellingly refer to this Great Aurora as "not natural" (Herndon and McGough 2008b: 272) and having come "too soon" (Herndon and McGough 2008b.: 273), unlike the fairy queen had intended, thereby potentially indicating how environmental changes are both caused by and have overpowered those who seek to control nature – despite simultaneous efforts of conservation.

After the terrible event, the fae are the only race on Lorwyn who retain their memories of a world unaffected by environmental catastrophe, which allows them to cast their intended version of Glen Elendra extended throughout Lorwyn as ecological utopia and to frame themselves as protectors of a natural balance between pastoral bliss and inevitable anthropocenic destruction. The final novel's ending highlights the omnipresence and omnipotence of *MTG*'s matriarch by showing that despite her apparent defeat by those seeking to hinder her reach on the realm, she cannot be overpowered. As part of the environment of the plane, Oona, just like nature itself, simply adapts to different realities and declares: "I am eternal. I am within the land, the rock, the rushing currents of air and water. My roots run deep and call to each other from afar." (Herndon and McGough 2008a: 308) The constant efforts of her children-as-subjects to pollinate their matron enable her to exude fertility and therewith self-procreate *as* nature, creating an entirely new generation of post-environmental catastrophe fae seeking to nourish Lorwyn and expand the lush vegetation of Glen Elendra once more (see Herndon and McGough 2008a: 309).

Taking into consideration the framing of Oona, Queen of the Fae, as both environmental protectress and power-hungry matriarch, *MTG*'s lore simultaneously offers multilayered comments on the impact of climate change and anthropocenic horrors on individuals, as well as on the interconnectedness of females and nature, potentially leading to either utopian or dystopian futures. While it is not the decided purpose of this chapter to relay the ways in which Oona both embodies female agency as well as outdated notions of the female as aligned with natural surroundings, procreation, and gossip (see, e.g. Herndon and McGough 2007: 84–5), it should be noted that as a popcultural twist on traditional fairy lore, *MTG*'s matriarch indeed provokes musings about the complex roles inhabited by females in postmodernity.

In line with Maria Tatar's perception of fairy tales, this particular adaptation of fairy lore challenges us "to identify differences and deviations" from earlier conceptualisations, "to wonder why the new departs from the old ... [and] what that reveals about our own cultural values" (2014b: 4). Comparing previous fictionalisations of fairies to the matriarch in question, a continuation of their inherent connectedness to nature is apparent, as well as a desire for cunning mischief. What this new fairy queen offers, though, is an immense investment in environmental protection, even at the expense of others. While it is certainly true that mythological belief in fairies has occasionally been the cause of socio-political decision-making (see Letcher 2001), *MTG*'s matriarch represents an even fiercer defence of her paradisiacal abode and her dreams of enlarging Glen Elendra's ecological utopia. Even though she is just one character among multitudes of wondrous creatures introduced by Wizards of the Coast, her decided play on established fairy lore casts Oona as a fictional force worthy of scrutiny – especially in light of contemporary environmental challenges caused by economically driven materialism. Drawing connections between Romantic and Victorian beliefs in fairies, Carole Silver argues that irrespective of the epoch in question, "all who asserted that fairies did exist did so with a sense that their reality was a protest against sterile rationality, evidence that the material and utilitarian were not sole rulers of the world" (1986: 148). Granted, a postmodern collectible trading card game like *MTG* perhaps does not expect its consumers to believe in the existence of their fictional characters, but the fact that a fairy holds such an esteemed place in Wizard of the Coast's universe shows both the important message conveyed by Oona specifically, as well as the relevance of games and game lore for contemporary storytelling and meaning-making. In the words of Jack Murray, "Magic has transformed and evolved into more than just a game. Magic has become a transmedia franchise that utilises its platform to engage players with its expansive storyworld. ... Magic: The Gathering exists as both a game platform and its own narrative transmedia property" (2023: 1).

Lesley Hawkes and Kelly Palmer

FernGully: The Last Rainforest (Bill Kroyer, 1992)

Fairy Eco-Warriors of a Living Australian Landscape

The 1992 family animation *FernGully: The Last Rainforest* is a US–Australian production that blends diverse fairy knowledges and folklore on an Australian natural landscape. The film is based on the book by Diana Young and the film's producer, her then-husband, Wayne Young. *FernGully* didactically champions environmental activism – also confronting concerns with industrialisation, commercialisation, and postcolonialism – but strikingly combines its multicultural environmentalism and lore with an eco-spirituality compatible with Australian Indigenous Knowledges. Despite the awkward lack of Indigenous Australian credits, the film is an eco-conscious allegory that celebrates the inherent spirituality of the natural Australian landscape and ultimately demonstrates the Indigenous Australian knowledge that "country speaks". Here, the fairies themselves are the embodiment of multi-cultures and landscapes – including Celtic, Anglo, and Hebrew knowledges, and Hollywood storytelling. Through these diverse lenses, the fairies implore the human character, Zak, as an avatar of the audience, to see themselves not as separate from the land but as one of its many expressions.

Australian "Fairies" as Eco-Warriors

The fairies, animals, humans, and spirits of *FernGully*, as well as the landscape itself, form an ensemble that represents a diverse ecosystem of harmonies and conflicts, though, not ones that need to be balanced but rather problems to be solved and absolved. The tension between humans and fairies and humans and the landscape drive the plot forward, as the fairies advocate and speak for the sovereignty of nature that the humans would industrialise. This tension is embodied in the character of Batty – a fruit bat voiced by Robin Williams – who appears in the film after having escaped an animal testing facility. With this implanted antenna that prompts him to repeat radio broadcasts and trauma-induced flashbacks, he speaks in a collage of discourses, blurring time and space, while he narratively and literally signals the oncoming presence of humans in the rainforest. Batty tells the fairies that he's been "brain-fried, electrified, infected and injectified / Vivosectified and fed pesticides". His descriptions of his time in a science lab frames his torture, while his schizophrenic monologue shows that he has become significantly less functional as a bat. As a result, the film shows that the dual embodiment of nature and industry/machine is in Batty's case not integrated, but results in pain and suffering. In this light, Batty is a non-consenting mediator between the rainforest and the machines, as well as an omen for what could befall the natural world if the deforestation machines continue to advance.

When the fairy protagonist Crysta meets the human, Zak, he is spray-painting trees with a red "X" and wearing a construction helmet. He listens to a portable music player with headphones, so divorcing himself from immersing with the landscape or from hearing the buzzsaws. When he spots Crysta's glow, he loses his music player, which is carried away in a creek, and trips on logs and is tangled in vines. When Crysta recites the spell, "Give the gift of fairy size" to save him from a felled tree, he falls into a spider's web on a tree destined for the woodchipper – so representing his thorough entanglement in the natural world and highlighting his kindred vulnerability to deforestation. It is also noteworthy that the landscape and Crysta work subconsciously but synchronistically in tandem to neutralise Zak and rearrange his position in the ecosystem, thus suggesting that Crysta and the landscape are different expressions of the same oneness.

FernGully: The Last Rainforest (Bill Kroyer, 1992)

Now seeing the world from his more diminutive stature, Zak is introduced and enchanted to the living rainforest: Crysta shows him how the plants and water respond or glow to his touch, and he must earn the trust of the fairies and animals who detest the humans who would destroy their home. The fairies are so connected with their environment that they can feel the pain of trees. After one of the trees is cut down, Crysta becomes distressed and says to Zak: "Can't you feel the tree's pain?". The human cannot understand and replies, "Its pain?" Having seen Zak recently heading into the woodchipper with the tree that he himself had marked, we are taught that the tree's pain and the human's pain are one and the same and our destinies intertwined. Throughout the film, there is an emphasis on the web-like connections between all species, and this is the message the fairies try to teach the human.

The fairies are thumb-sized, petite, jagged-skirt wearing, winged humanoids in the style of Tinkerbell. They unanimously have a purplish-grey skin tone that is a few shades lighter than the blonde Zak's tan (a stereotypical representation of ideal Australian masculinity), and have different hair colours from black to vibrant orange. It is clear that the fairies are not supposed to have a race or ethnicity, though given their ambassadorship of the Australian landscape, their universality seems rather like a Euro-American import. Figures of fairies in Australian culture are predominantly of Anglo-Celtic origin, as are 58 per cent of the Australian population (Australian Human Rights Commission 2018: 7). Crucially, Australia lacks "an official national folklore register", and folklore "is academically trivialised and generally misunderstood, yet it has the potential to ease social tensions and heal cultural rifts" (McCarron 2013: para. 8).

Little People: Beyond Binaries

However, Australian Indigenous Knowledges document very different kinds of fairies, or fae-like spirits. Little People share company with Australian creature-spirits, such as the famed bunyips and yowies, though they have more in common with, and indeed are sometimes synonymous with, Mimi [mimih], "tall, thin beings that live in the rocky escarpment of northern Australia as spirits" (Australian Museum 2019: para. 7–8). Little People bear physical and supernatural similarities with various types of Anglo-Celtic fairies, elves, dwarves, and leprechauns, while Varner classifies the "tiny little men" as "nature spirits" in a similar kingdom as fairies (2007: 34). Names for Little People vary across Australian languages and countries, though descriptions of Little People as "standing knee high, pungent in smell and hairy in appearance … [and] ugly" are "strikingly consistent among Aboriginal cultures" (Lee-Ryder and Roman 2018: 9). Little People's physical strength, spiritual advancement, and immersion with the natural Australian landscape enshrine them with both reverence and fear; they are metonyms for nature's mischievous power striking a balance between life and death. Indeed, they can be healing guardians or can "steal your soul" (Heenan quoted in Lee-Ryder and Roman 2018: 9), and are shapeshifters able to turn invisible and transform, blurring the boundaries between plant, animal, spirit, and light. But territory remains important: "if you are harming or stealing from land then you may face the wrath of the little people. They are guardians of protectors of the land" (Lee-Ryder and Roman 2018: 9).

This complex role also lends Little People to be read as autochthonous – as physical embodiments for how *country speaks*. The Indigenous Knowledge of a country that speaks refers to the way that people, plants, and even language are borne from the earth itself. That is, everything on the planet feeds, grows, and consumes everything else on the planet, making everything part of the same source and cycle, while language itself is believed to be constant, breathed out of the earth and taught to those willing to listen. Although *FernGully*'s "Australia" fairies resonate with a global familiarity rather than with Indigenous Knowledges, they are depicted as nature spirits and serve a similar purpose of protecting the land. However, *Ferngully*'s fairies are not here to maintain or restore a balance between light and shadow, growth and destruction, but rather neutralise any threat. Hence, the film ultimately shows that for all living creatures to live in harmony as one family and ecosystem, there are binaries to first be dissolved, most notably between the side of conscious coexistence (fairies, native animals, and rainforest) and the consumers (humans, industrialisation, the evil spirit Hexxus).

Related binaries of native/intruder, colonised/coloniser, consumed/consumer, nature/industry, good/evil also speak through metaphors, and so the postcolonising aspect of the film unfolds. Aileen Moreton-Robinson (2003) gives us the term "postcolonising" to highlight the ongoing appropriation and erasure of Australian Aboriginal people and culture, which thus remains in the present tense. Awkwardly, the film metaphorically confronts the colonisers (white humans) as the deforestation industry dispossesses the fairies, animals, and plants; however, the film itself seems to substitute Euro-American myth, characters, and producers instead of acknowledging the sovereignty or presence of Australia's Aboriginal people. It is unclear how much of the whitewashing and Americanisation of the film is referenced consciously, as Batty, a native Australian animal, is voiced by a US actor and is said to dance "like Elvis" and sometimes spits out French words. Indeed, he is an example of how globalisation silences and scrambles native Australian voice.

This is most glaring in the opening of the film, where Indigenous Australian-style art shows stick figure-people and native Australian animals. The camera pans over Indigenous-style art depicting a human figure holding a blue fairy-figure in the palm of their hand. Magi narrates, "the closest of our friends were humans", but when a volcano (Wollumbin) releases the destructive spirit Hexxus, it is the humans who eventually give him the vehicle (literally and metaphorically) to carry out his consumption of the forest. The relationship between fairies (as active agents of the landscape) and humans (as the conquerors of that landscape) functions not only as an environmental allegory but also as a metonym for Australian colonisation and the burying of Indigenous Knowledges. The film gives the sense that humans and fairies must become friends again to save the land, but from a postcolonial perspective, the film is also implying that Indigenous Knowledges must be brought into the mainstream and that Aboriginal Australians and settler Australians must reconcile with this knowledge. As a result, the fairies occupy an awkward role of advocating but substituting for an Indigenous presence in their roles as environmental activists. This whitewashing is common across colonial Australian fairy stories, where it is the landscape itself that makes anything "Australian".

Storied Land

The fern gully of the film is fictional in the sense that it is not actually taken from a particular mapped area, yet like the earlier fairy stories of non-Indigenous Australia, the landscape is paramount in *FernGully*. As previously discussed, the representation of the physical fairy was imported from the European tradition; colonial Australian fairy stories do not include new representations of fairies, elves, or sprites but their newness stemmed from the use of setting for the fairies rather than the fairies themselves:

> The uniqueness of the Australian landscape and its flora and fauna – from the European perspective – allowed the actual environment to fulfill the role of the fairy realm, but the fairies themselves emigrated from England and arrived in this realm in their European diaphanous gowns and fashionable hair (Rozario 2011: 14).

Alongside this is a spiritual dimension to the rainforest of *Ferngully* that draws on Australian Indigenous as well as Celtic, Anglo, and global environmentalities. In the opening sequence of the film, an elder fairy named Magi calls her people "tree spirits" and explains that when "the forest went on forever", before human industrialisation, the tree spirits "nurtured the harmony of all living things". The fairies are set up as having always lived on this land and belonging to nature; they are an integral part of the landscape. Magi is seen as having the strongest connection with the land around her to the point where the landscape comes to life as she passes. She has been there since "the very first tree" (4) and promises to teach Crysta the "Old Powers, for knowledge must never be lost" (4). It is unclear what this knowledge is, but it does highlight that there are other forms of knowledge existing in and of the land. This representation of autochthony echoes the Indigenous Australian knowledge that "country speaks", that knowledge and language itself is indigenous to a physical geography. Thus, the fairies' embodiment of Australian Indigenous Knowledges is in many ways ironic, since the fairies themselves seem transplanted onto the landscape, and yet their lessons teach that connection to the landscape is determined only by conscious participation. The ability for even the most "toxic" of characters – from the humans to the destructive spirit, Hexxus – to reintegrate with the land shows that the binary between nature and industry, consumed and consumer, is an illusion, and that all are indeed part of one living landscape.

Healing the "Toxic"

The film provides an evil villain in the form of Hexxus – the dark manifestation of oil, poisonous sludge, and toxic waste – but the humans are the consumers to be reformed. It is Zak who accidentally releases the villain from his prison in a large Baobab tree accidentally spraying an X on it. The tree is cut down releasing Hexxus. The Australian Baobab tree (bottle tree or larrgadiy tree) is usually found in Western Australia and Northern Territory but can be found across Australia, and these trees can grow huge over 1,500 years. Hence, to cut one down could be seen as sacrilegious and ignorant of the rainforest's delicate and long-grown ecosystem. What is released, then, is Hexxus as the embodiment of human greed, selfishness, and wilful ignorance: a curse to be healed. He feeds and grows off pollution and his sole aim is to destroy the environment and the fairies: "I suck them dry", he sings.

However, Hexxus, voiced by Tim Curry, shifts our understanding of environmental villains. His tantalising version of the song "Toxic Love" highlights how alluring industry and financial gain can be. Humans may not have invented Hexxus, but their greed allows him to grow: "greedy human beings will always lend a hand". Curry's singing somehow manages to make acid rain sexy and forgivable, inviting the audience to see their resonance with its "evil". One of the strengths of the film is its ability to reveal the dangers of complacency, and it is the fairies who teach Zak how to enact his ethical, environmental filter. Hexxus is defeated by a tree seed that Crysta plants inside his core, which overtakes his power. As he shrinks he shouts, "what happened to the energy?" The answer, of course, is that it is in the tree and the community of fairies. Thus, the binary of growth and destruction is not held in balance but resolved by transmuting Hexxus's power into a regenerative force so that "Hexxus can never harm Fern Gully again". However, Zak acknowledges that "humans still could".

Indeed, that Hexxus was not created by humans, but rather born out of a volcano, shows that Hexxus is not an introduced threat to the landscape such as an invasive coloniser but is part of the landscape to be brought back into harmony. Because the humans evacuated the region leaving the fairies to deal with this disastrous creation, and because humans are now responsible for releasing and "feeding" Hexxus, Zak plays a part in this resolution in a symbolic repayment of karma. Interestingly, the name Zak, which sounds pixie-like on its own, comes from the Hebrew prophet Zecheriah, which means "Yahweh remembers". In the film's finale Zak tells Crysta that he must "go back" to ensure that humans understand their own toxicity. Crysta gives Zak a seed – a parallel to the one that transformed Hexxus – that she enchants to glow blue, and says, "Remember Zak; remember everything". Here, both Zak and Hexxus have their transformation out of destruction and into harmony marked by a seed. Zak is then returned to his human size and quickly finds a human colleague: it is implied here that Zak goes forth as a kind of prophet to advocate for the life and sovereignty of the living landscape, and all earthly beings' stake in its life. This gives the fairy story a kind of biblical didacticism, and in the idea that "God" (nature) speaks through the fairies and Zak, connecting all these beings and consciousnesses as one voice. This is reminiscent again of the idea that *country speaks*, wherein the spirit of the landscape itself is the implied voice of the film, which implores Zak and the audience together to "remember" that our fates are the same.

Concluding Thoughts

The fairies of *Ferngully* recite Dreaming-like stories; defend the native plants and animals against deforestation and pollution; and fight for the eco-consciousness of the human character against the maleficent smoke monster spirit manifested from the fumes and oil of construction machinery. The fairies and native Australian animals living on and protecting the natural or "uncolonised" Australian rainforest stand off against the colonial oppressors – who, in the name of science and industry, log and pollute the forest and abuse the animals. However, the binaries of nature/industry, organic/pollution, and hero/villain are not stable in this film, as both Hexxus and Zak show that the colonising industrial world and the natural world are one ecosystem, and that thresholds between the human and fairy worlds are simply a matter of perspective.

Although the imported fairies of *FernGully* do little to represent Indigenous Australian knowledge, their role as ambassadors of the land echoes a spiritual environmentalism that is mostly compatible with Indigenous Knowledge; however, while the "Little People" of Australia on the landscape draws attention to one's behaviours and effects on the environment, and, indeed, are positioned in the ecosystem as guardians of the landscape, they are not moralising figures: instead, the nature spirits of Indigenous Australia seek a balance of growth and destruction rather than a resolution of pure harmony, as is seen in *FernGully*.

Fairy of the Forest

Gemma Files

Bibliography

A.I. Artificial Intelligence, dir. Spielberg, Steven (DreamWorks, 2001).
Aaronovitch, Ben, and Andrew Cartmel, *Rivers of London: The Fey and the Furious* (London: Titan Comics, 2019–2020).
Aaronovitch, Ben, *Rivers of London* (London: Gollanzc, 2011b).
———, *Moon over Soho* (London: Gollancz, 2011a).
———, *Whispers Under Ground* (London: Gollancz, 2012).
———, *Broken Homes* (London: Gollancz, 2013).
———, *Foxglove Summer* (London: Gollancz, 2014).
———, *Lies Sleeping* (London: Gollancz, 2018).
———, *What Abigail Did That Summer* (London: Gollancz, 2021).
Ackroyd, Peter, *Albion: The Origins of the English Imagination* (London: Chatto and Windus, 2002).
Adamson, Walter L., *Hegemony and Revolution: A Study of Antonio Gramsci's Political and Cultural Theory* (Berkeley and Los Angeles, CA: University of California Press, 1980).
Agnew, John, "Landscape and National Identity in Europe: England versus Italy in the Role of Landscape in Identity Formation", in, Zoran Roca, Paul Claval, and John Agnew, eds., *Landscapes, Identities and Development* (London: Routledge 2016), 37–50.
Ahlawat, Neerja, "Dispensable Daughters and Indispensable sons: Discrete Family Choices", *Social Change* 43 (2013), 365–7.
———, "'The Political Economy of Haryana's Khaps", *Economic and Political Weekly* 47 (2012), 15–17.
Ahn, Somi, "National Regeneration Through Childhood in Edith Nesbit's *The Story of the Amulet*", *Children's Literature in Education*51 (2020), 348–60.
Alberghini, Jennifer, "Matriarchs and Mother Tongues: The Middle English *Romans of Partenay*", in Misty Urban, Deva F. Kemmis, and Melissa Ridley Elmes eds., *Melusine's Footprint: Tracing the Legacy of a Medieval Myth* (Leiden and Boston: Brill, 2017).
Albury, William R., "From Changelings to Extraterrestrials: Depictions of Autism in Popular Culture", *Hektoen International: A Journal of Medical Humanities* 3 (2011), <https://hekint.org/2017/01/30/from-changelings-to-extraterrestrials-depictions-of-autism-in-popular-culture/>, accessed 11 March 2024.

Aldiss, Brian, "Super-Toys Last All Summer Long" *Wired* (January 1997), <www.wired.co.uk/1997/01/ffsupertoys>, accessed 14 August 14 2022.

Alexander, Lloyd, *The Book of Three* [1964] (New York: Henry Holt and Company, 2007).

Alexander, Skye, *Fairies: The Myths, Legends, and Lore* (New York: Simon & Schuster, 2014).

Arcus, Doreen, "Vulnerability and Eye Colour in Disney Cartoon Characters", in J. Steven Reznick, ed., *Perspectives on Behavioral Inhibition* (Chicago: The University of Chicago Press, 1989), 291–8.

Aurell, Martin, "Henry II and Arthurian Legend", in Christopher Harper-Bill and Nicholas Vincent, eds., *Henry II: New Interpretations* (Woodbridge: Boydell Press, 2007), 362–94.

Australian Human Rights Commission, "Leading for Change: A Blueprint for Cultural Diversity and Inclusive Leadership Revisited" (2018), <https://humanrights.gov.au/sites/default/files/document/publication/Leading%20for%20Change_Blueprint2018_FINAL_Web.pdf>, accessed 18 June 2023.

Australian Museum, "Dreaming Stories" (2019), <https://australian.museum/about/history/exhibitions/indigenous-australians/>, accessed 18 June 2023.

Baird-Jackson, Jason, "On Cultural Appropriation", *Journal of Folklore Research* 58 (2021), 77–122.

Baker, David, Stephanie Green, and Agnieszka Stasiewicz-Bieńkowska, eds., *Hospitality, Rape and Consent in Vampire Popular Culture: Letting the Wrong One In* (Cham: Palgrave Macmillan, 2017).

Ballard, Linda-May, "Fairies and the Supernatural in the Reachrai", in Peter Narvaez, ed., *The Good People: New Fairylore Essays* (Lexington: University Press of Kentucky, 1991).

Ballard, Linda-May. "A Singular Changeling?", *Folk Life* 52/2 (2014), 137–51.

Bane, Theresa, *Encyclopedia of Fairies in World Folklore and Mythology* (Jefferson: McFarland & Company, 2013).

Barbie. dir. Greta Gerwig (Warner Bros Pictures, 2023).

Barnes, Brooks, "Disney Hoping 'Tinker Bell' Spreads Fairy Dust on Sales", *New York Times* B7(L) (31 October 2008).

Barnett, P. Chad, "Reviving Cyberpunk: (Re)Constructing the Subject and Mapping Cyberspace in the Wachowski Brothers' Film *The Matrix*", *Extrapolation* 41/4 (2000), 359–74.

Barrie, J. M, *Peter and Wendy* (New York: Barnes & Noble, Inc., 2014).

Barros, David, "Walt Disney's Peter Pan", *e-fabulations* (2007), <https://ler.letras.up.pt/uploads/ficheiros/4291.pdf>, accessed 18 June 2023.

Bates, Thomas R., "Gramsci and the Theory of Hegemony", *Journal of the History of Ideas* 36/2 (1975), 351–66.

Bavidge, Jenny, "Exhibiting Childhood: E. Nesbit and the Children's Welfare Exhibitions", in Adrienne E. Gavin and Andrew F. Humphries, eds, *Childhood in Edwardian Fiction: Worlds Enough and Time* (New York: Palgrave Macmillan, 2009), 125–42.

Bell, Elizabeth, "'Do You Believe in Fairies?' Peter Pan, Walt Disney, and Me", in Douglas Brode and Shea T. Brode, eds., *It's the Disney Version!: Popular Cinema and Literary Classics* (Lanham: Rowman & Littlefield, 2016), 79–91.

Benjamin, Walter, and Asja Lacis, "Naples" in Peter Demetz, ed., *Reflections: Essays, Aphorisms, Autobiographical Writings* (New York: Schocken, 1978).

Berard, Christopher Michael, *Arthurianism in Early Plantagenet England: From Henry II to Edward I* (Woodbridge: Boydell Press, 2019).

Binney, Sara Helen, "How 'the Old Stories Persist': Folklore in Literature after Postmodernism", in *C21 Literature: Journal of 21st-Century Writings* 6/2 (2018), 1–20.

Bisset, Jennifer, "Creator of The Matrix code reveals its mysterious origins", CNET (19 October 2017).<https://www.cnet.com/culture/entertainment/lego-ninjago-movie-simon whiteley-matrix-code-creator/>, accessed 10 December 2023.

Blackford, Holly, "PC Pinocchios: Parents, Children and the Metamorphosis Tradition in Science Fiction", in Sharon R. Sherman and Mikel J. Koven, eds, *Folklore/Cinema: Popular Film as Vernacular Culture* (Logan: Utah State University Press, 2007). 74–92.

Bone, Kristin L., "Murders Most Foul: Changing Myths", in Helen Gavin, ed., *Women and the Abuse of Power: Interdisciplinary Perspectives* (Bradford: Emerald Publishing Limited, 2022) 31–42.

Borowska-Szerszun, Sylwia, "Ethnic and Cultural Diversity in Ben Aaronovitch's Urban Fantasy Cycle Rivers of London", *Nordic Journal of English Studies* 18/1 (2019), 1–26.

Bosacki, T. K., "Tinker Bell meet and greet still closed as pixie is labelled 'problematic'", in *DisneyFanatic*, 20 April 2022. <https://www.disneyfanatic.com/tinker-bell-meet-and-greet-unavailable-as-pixie-is-labeled-problematic-tb1/tb1/#:~:text=While%20it%20is%20not%20confirmed,jealous%20of%20Peter%20Pan's%20attention.%E2%80%9D>, accessed 18 June 2023.

Botting, Fred, *Gothic* (Abingdon; New York: Routledge, 2014).

Bourke, Angela, "Reading a Woman's Death: Colonial Text and Oral Tradition inNineteenth-Century Ireland" in *Feminist Studies* 21/3 (1995), 553–86.

Bourke, Angela, *The Burning of Bridget Cleary* (London: Pimlico, 1999).

Bown, Nicola, *Fairies in Nineteenth-Century Art and Literature* (Cambridge: Cambridge University Press, 2001).

Branford, Anna, "Gould and the fairies", *The Australian Journal of Anthropology*, 22 (2011), 89–103.

Brewer, Derek, "The Colour Green", in Derek Brewer and Jonathan Gibson, eds., *The Companion to the* Gawain-*poet* (Cambridge: DS Brewer, 1997), 181–90.

Breytenbach, Ria Mariza, "Recasting the social critic: social commentary in selected novels of Charles Dickens and Terry Pratchett", PhD dissertation (North-West University, Potchefstroom Campus, 2016).

Briggs, Katharine M., "The English Fairies", *Folklore* 68/1 (1957), 270–87.

———, "Making a Dictionary of Folk-Tales", *Folklore* 72/1 (1961), 300–5.

———, *The Fairies in Tradition and Literature* (London: Routledge, 1967).

———, *The Fairies in Tradition and Literature* (London: Routledge & Kegan Paul, 1968).

———, "The Fairies and the Realms of the Dead." *Folklore* 81/2 (1970), 81–96.

———, *A Dictionary of Fairies* (London: Allen Lane, 1976).

———, *The Vanishing People: A Study of Traditional Fairy Beliefs* (London: B.T. Batsford, 1978).

———, *The Fairies in Tradition and Literature* [1967] (London: Routledge, 2002).

Broderick, Shane, "The Evolution of the Irish Otherworld", *Ireland's Folklore and Traditions*, (26 July 2019). <https://irishfolklore.wordpress.com/2019/07/26/the-evolution-of-the-irish-otherworld/>, accessed 11 March 2024.

———, "Folklore, Story, and Place: An Irish Tradition with Vast Touristic Value", in Jack Ironside and Rachael Hunter, eds, *Folklore, People and Place: International perspectives on tourism and tradition in storied places* (Routledge, Taylor and Francis, 2023).

Brophy, Christina. "'Her own and her children's share': luck, misogyny and imaginative resistance in twentieth-century Irish folklore", *Irish Historical Studies* 46/169 (2022), 155–78.

Buckley, Arabella B., *The Fairy-Land of Science* (Charing Cross: Edward Stanford, 1880).

Bullivant, Richard, *Parallel Worlds and the Existence of Fairies* (London: Create Space Independent Publishing, 2017)

Burnett, Frances Hodgson, *A Little Princess* (Hertfordshire: Wordsworth Editions Limited, 1994).

Burrow, Merrick, "The Cottingley Fairies: A Study in Deception", *Leeds University Library Galleries/Good Arts & Culture*, September 2020. <https://artsandculture.google.com/story/the-cottingley-fairies-a-study-in-deception-leeds-university-library-galleries/lgVB6Ceti9WVAw?hl=en>, accessed 1 August 2023.

Bynum, Carol W., "Material Continuity Personal Survival and the Resurrection" in *History of Religions* 30/1, The Body (August 1990). 51–85.

Campbell, Harry, *Supernatural Scotland* (Glasgow: HarperCollins, 1999).

Carey, John, "Time, Space, and the Otherworld", *Proceedings of the Harvard Celtic Colloquium* 7 (1987), 1–27.

———, "Otherworlds and Verbal Worlds in Middle Irish Narrative", *Proceedings of the Harvard Celtic Colloquium* 9 (1989), 31–42.

Carpenter, Alexander, "The 'ground zero' of goth: Bauhaus, 'Bela Lugosi's Dead' and the origins of gothic rock", *Popular Music and Society* 35/1 (2012), 25–52.

Census of India, Haryana, 2001 and 2011. Director of Census operations, Haryana.

Chambers, Dewey, "The Disney Touch and the Wonderful World of Children's Literature", *University of the Pacific, Scholarly Commons*, 1 January 1966. <https://scholarlycommons.pacific.edu/dewey-chambers/3>, accessed 18 June 2023.

Chaplin, Susan, *The Postmillennial Vampire: Power, Sacrifice and Simulation in True Blood, Twilight and Other Contemporary Narratives* (Cham: Palgrave Macmillan, 2017).

Cherry, Brigid, *True Blood: Investigating Vampires and Southern Gothic* (New York: I.B. Tauris, 2012).

Cinderella Story, A. dir. Mark Rosman (Warner Bros. Pictures, 2004).

Clark, John. *The Green Children of Woolpit*. Exeter: University of Exeter Press, 2024.

Clarke, Philip A., "Indigenous Spirit and Ghost Folklore of "Settled" Australia". *Folklore* 112/2 (2007), 141–61.

Clarke, Susanna, *Jonathan Strange and Mr Norrell* (London: Bloomsbury, 2004).

Colfer, Eoin, *Artemis Fowl* (London: Viking, 2001).

Collins, Kira. "Non-Traditional Motherhood in Contemporary Irish Film: Carmel Winters' Feature Film Snap (2010) and Her Short Film Limbo (2008)", *Athens Journal of Humanities & Arts* 7/1 (2020), 85–104.

Collodi, Carlo, *Pinocchio: The Story of a Puppet* (Philadelphia: J.B. Lippincott Company, 1916).

Connolly, John, *Nocturnes* (London: Hodder, 2004).

———, "The New Daughter" in *Nocturnes* (London: Hodder, 2004), 107–22.

Conrad, JoAnn, "Changeling", in Donald Haase, ed., *The Greenwood Encyclopedia of Folktales and Fairy Tales* (Westport, CT: Greenwood Press, 2008), 179–80.

Cooper, Helen, *The English Romance in Time: Transforming Motifs from Geoffrey of Monmouth to the Death of Shakespeare* (Oxford: Oxford University Press, 2004).

Cooper, Lucy, *The Element Encyclopaedia of Fairies* (London: Harper Collins, 2014).

Crafton, Donald, "Animation: The Last Night in the Nursery: Walt Disney's 'Peter Pan'", *Velvet Light Trap* 24 (Fall 1989), 33–52, <https://login.ezproxy.une.edu.au/login?url=https://www.proquest.com/scholarly-journals/animation-last-night-nursery-walt-disneys-peter/docview/740767601/se-2>, accessed 18 June 2023.

Crawley, Geoffrey, "That Astonishing Affair of the Cottingley Fairies: Part One", *The British Journal of Photography* 129 (24 December 1982), 1374–80.

———, "That Astonishing Affair of the Cottingley Fairies: Part Two", *The British Journal of Photography* 129 (31 December 1982), 1406–14.

———, "That Astonishing Affair of the Cottingley Fairies: Part Six", *The British Journal of Photography* 130 (4 February 1983), 117–24.

Croft, Janet Brennan, "Nice, good, or right: faces of the wise woman in Terry Pratchett's witches' novels", *Mythlore* 26, no.3/4 (2008), 151–64.
Cunliffe, J. W., "The Queenes Majesties Entertainment at Woodstocke", *Proceedings of the Modern Language Association* 26/1 (1911), 92–141.
Daimler, Morgan, *A New Dictionary of Fairies: A 21st Century Exploration of Celtic and Related Western European Fairies* (London: Collective Ink, 2020).
———, "Pointed Ears", in *A New Dictionary of Fairies* (Winchester: Moon Books, 2020), 267–71.
———, *Pagan Portals: Aos Sídhe, Meeting the Irish Fair Folk* (Berkeley: Moon Books, 2022).
Daisy Chain, The, dir. Aisling Walsh (Content Film, 2008).
Davis, Janet, M., "Introduction", in Tiny Kline, *Circus Queen and Tinker Bell: The Memoir of Tiny Kline* (Champaign: University of Illinois Press, 2008).
Deffenbacher, Kristina, "Hybrid Heroines and the Naturalization of Women's Violence in Urban Fantasy Fiction", in Amanda Hobson and U. Melissa Anyiwo, eds., *Gender in the Vampire Narrative* (Rotterdam: Sense Publishers, 2016), 29–44.
Dholakia, Nikhilesh and Jonathan Schroeder, "Disney: Delights and Doubts", *Journal of Research for the Consumer* 1/2 (2001). <http://jrconsumers.com/Academic_Articles/issue_2/DholakiaSchroeder.pdf>, accessed 18 June 2023.
Dinello, Daniel, *Technophobia!: Science Fiction Visions of Posthuman Technology* (Texas: Texas University Press, 2005).
"Disney in Tinker Bell rights deal", *Marketing* (29 June 2005).
Dodge, Autumn M. with Paul A. Crutcher, "Examining Literacy Practices in the Game Magic: The Gathering", *American Journal of Play* 10/2 (2018), 168–92.
Donohue, Keith, *The Stolen Child* [2006] (London: Vintage, 2007).
Doyen, Eugene, *Sookie Stackhouse and Pierre Bourdieu* (N.L: Lulu.com, 2014).
Doyle, Arthur Conan, *The New Revelation* (London: Hodder & Stoughton, 1918).
———, "Fairies Photographed: An Epoch-Making Event", *The Strand Magazine* (December 1920).
———, *The Coming of the Fairies* (London: Hodder & Stoughton, 1922).
———, "The Adventures of the Devil's Foot", in *Sherlock Holmes: The Complete Novels and Stories* (New York: Bantam Dell, 2003).
"E. Nesbit", *Encyclopaedia Britannica*, 30 April 2023. <https://www.britannica.com/biography/E-Nesbit>, accessed 30 April 2023.
Eager, Edgar, "Daily Magic", *The Horn Book Magazine* (1958), 348–58.
Eason, Cassandra, *The Magick of Faeries: Working with the Spirits of Nature* (Minnesota: Llewellyn, 2013).
Eberley, Susan, "Fairies and the Folklore of Disability: Changelings, Hybrids and the Solitary Fairy", *Folklore* 99/1 (1988), 58–77.

Eberly, Susan Schoon, "Fairies and the Folklore of Disability: Changelings, Hybrids and the Solitary Fairy", *Folklore* 99/1 (1988), 58–77.

Ekman, Stefan, "Urban Fantasy: A Literature of the Unseen", *Journal of the Fantastic in the Arts* 27/3 (2016), 452–69.

Ella Enchanted, dir. Tommy O'Haver (Miramax, 2004).

England, Dawn Elizabeth, Descartes, Lara and Collier-Meek Melissa A., "Gender Role Portrayal and the Disney Princesses" in *Sex Roles* 64 (2011) 555–67.

Enstone, Zoë, "'Wichecraft & Vilaine': Morgan Le Fay in Medieval Arthurian Literature", PhD dissertation (University of Leicester, 2011).

Espinoza Garrido, Felipe, "Queerness in the Neo-Victorian Empire: Sexuality, Race, and the Limits of Self-Reflexivity in Carnival Row and The Terror", *Neo-Victorian Studies*, 13/1 (2020), 212–41.

Espírito Santo, Diana, and Ruy Blanes, "Introduction: On the Agency of Intangibles", in R. Blanes and D. Espírito Santo, eds, *The Social Life of Spirits* (Chicago: University of Chicago Press, 2014).

Evan Torner, and Jonathan Walton, "Tabletop Role-Playing Games", in José P. Zagal and Sebastian Deterding, eds, *Role-playing Game Studies: Transmedia Foundations* (New York and London: Routledge, 2018), 63–86.

Evans-Wentz, W.Y., *The Fairy Faith in Celtic Countries* (London: Henry Frowde, 1911).

Excalibur, dir. John Boorman (Warner Bros., 1981).

"Extraordinary and Calamitous Occurrence Near Beaufort", *Kerry Sentinel* (31 January 1888).

"Extraordinary Murder Near Killarney", *Weekly Irish Times* (4 February 1888).

"Fairy", *Oxford English Dictionary Online*, December 2021. <https://www-oed-com.bibproxy.kau.se/view/Entry/67741?redirectedFrom=fairy#eid>, accessed 26 January 2022.

Faludi, Susan, *Backlash: The Undeclared War Against American Women* (New York: Random House LLC, 2006).

Femia, Joseph V., *Gramsci's Political Thought: Hegemony, Consciousness, and the Revolutionary Process* (Oxford: Oxford University Press, 1981).

Ferngully: The Last Rainforest, dir. Brian Kroyer (20th Century Fox, 1992).

Fischer, Lucy, *Cinematernity: Film, Motherhood, Genre* (Princeton, NJ: Princeton Legacy Library, 2016).

Flannery-Dailey, Frances, "Robot Heavens and Robot Dreams: Ultimate Reality in A.I. and Other Recent Films", *Journal of Religion and Film* 7/2 (2003), 1–36.

Forsberg, Laura, "Nature's Invisibilia: The Victorian Microscope and the Miniature Fairy", *Victorian Studies* 57/4 (2015), 638–66.

Foster, Michael Dylan, "Introduction: The Challenge of the Folkloresque," in Michael Dylan Foster and Jeffrey A. Tolbert, eds, *The Folkloresque: Reframing Folklore in a Popular Culture World* (Boulder: Utah State University Press, 2016), 3–33.

Foucault, Michèl, *The Order of Things: An Archaeology of the Human Sciences* (New York: Vintage Books, 1994).
——, *Archaeology of Knowledge* (Abingdon, New York: Routledge, 1995).
Frair Rush and the Frolicsome Elves", *Littell's Saturday Magazine: A Collection of Light Reading* 1 (Jul-Dec 1836), 517–21.
Frazer, Sir James George, *The Golden Bough: A Study in Magic and Religion* (Oxford: Oxford University Press, 1998).
Frost, William Henry, *Fairies and Folk of Ireland* (London: DigiCat Books, 2022).
Gaiman, Neil, "Terry Pratchett isn't Jolly. He's Angry", *The Guardian*, 24 September 24 2014. <https://www.theguardian.com/books/2014/sep/24/terry-pratchett-angry-not-jolly-neil-gaiman#:~:text=Terry%20looked%20at%20me.,knew%20that%20he%20was%20right>, accessed 10 July 2021.
Galvan, Pilar, "*Nanny* employs African folklore in a haunting Black horror film". *WFAE – NPR*, 23 November 2022. <https://www.wfae.org/2022-11-23/nanny-employs-african-folklore-in-a-haunting-black-horror-film>, accessed 27 December 2023.
Gavin, Adrienne E., and Andrew F. Humphries, "Worlds Enough and Time: The Cult of Childhood in Edwardian Fiction", in Adrienne E. Gavin and Andrew F. Humphries, ed., *Childhood in Edwardian Fiction: Worlds Enough and Time* (New York: Palgrave Macmillan, 2009), 1–20.
Gilligan, Cecily, *Cures of Ireland: A Treasury of Irish Folk Remedy* (Newbridge: Merrion Press, 2023).
Goldring, Elizabeth, "Portraiture, Patronage, and the Progresses: Robert Dudley, Earl of Leicester, and the Kenilworth Festivities of 1575", in Jayne Elizabeth Archer, Elizabeth Goldring, and Sarah Knight, eds, *The Progresses, Pageants, and Entertainments of Queen Elizabeth I* (New York: Oxford University Press, 2007), 163–88.
Golub, Alex, and Jon Peterson, "How Mana Left the Pacific and Became a Video Game Mechanic", in Matt Tomlinson and Ty P. Kawika Tengan, eds, *New Mana. Transformations of a Classic Concept in Pacific Languages and Cultures* (Canberra: ANU Press, 2016), 309–48.
Gomes, Luce, K. C. Pereira, Joyce Christianne, João Lucas Campos, and Karl Spracklen, "Light, camera, hospitality: relationship between hosts and guests in film productions in Minas Gerais, Brazil" in *Leisure Studies* (2023), 1–13.
Gramsci, Antonio, *Prison Notebooks: Volume I*, Joseph A. Buttigieg ed., and Joseph A. Buttigieg and Antonio Callari trans. (New York: Columbia University Press, 1992).
Gravelle, Cody, "Magic: The Gathering's Next Set Brings Back Faeries," *Screenrant*. Screenrant.com, 18 July 2019. <https://screenrant.com/magic-gathering-faeries-throne-eldraine-new-set/>, accessed 20 June 2023.
Green, Carolyn Eve and Edmund Lenihan, *Meeting the Other Crowd* (Basingstoke: Gill and McMillan Ltd, 2003).

Green, Richard, *Elf Queens and Holy Friars: Fairy Beliefs and the Medieval Church* (Philadelphia: University of Pennsylvania Press, 2016).
Greenblatt, Stephen, *Renaissance Self-Fashioning: From More to Shakespeare* (Chicago and London: University of Chicago Press, 1980).
Grelle, Bruce, *Antonio Gramsci and the Question of Religion: Ideology, Ethics, and Hegemony* (London and New York: Routledge, 2017).
Guiley, Rosemary, *Fairies* (New York: Chelsea House, 2010).
Haapala, Arto, *The City as Cultural Metaphor: Studies in Urban Aesthetics* (Helsinki: International Institute of Applied Aesthetics, 1998).
Hahn, Daniel, "Fairies", in *The Oxford Companion to Children's Literature* (Oxford: Oxford University Press, 2015), n.p.
Harf-Lancner, Laurence, *Les fées au Moyen Âge: Morgane et Mélusine; La naissance des fées* (Geneva: Editions Slatkine, 1984).
Harte, Jeremy. Fairy Encounters in Medieval England. Exeter: University of Exeter Press, 2024.
Hartland, Sidney, *The Science of Fairy Tales: An Enquiry into Fairy Mythology* (London: DigiCat Books, 2022).
Healy, Millie Mae, "'Fate: The Winx Saga' Doesn't Know Why Anyone Liked 'Winx Club': Season Review", *The Harvard Crimson*, 29 January 29 2021. <https://www.thecrimson.com/article/2021/1/29/fate-the-winx-saga-season-1-review/>, accessed 11 March 2024.
Hebert, Jill M., *Morgan Le Fay, Shapeshifter* (New York: Palgrave Macmillan, 2013).
Heffernan, Teresa, "A.I. Artificial Intelligence: Science, Fiction and Tales", *English Studies in Africa* 61/1 (2018), 10–15.
Hegarty, Mandy, "Interview: Irish Writer/Director Aisling Walsh Shares Secrets of Filmmaking", IFTN, 3 November 2011. <http://www.iftn.ie/production/news/?act1=record&only=1&aid=73&rid=4284411&tpl=archnews&force=1>, accessed 11 March 2024.
Henderson, Lizanne, and Edward J. Cowan, *Scottish Fairy Beliefs: A History* (Glasgow: Tuckwell Press, 2001).
Hendrix, Grady, *Paperbacks from Hell* (Philadelphia: Quirk Books, 2017).
Hennard Dutheil de la Rochère, Martine, Gillian Lathey, and Monika Wozniak, "Introduction: Cinderella across Cultures. Cinderella in the Twenty-First Century", in Martine Hennard Dutheil de la Rochère, Gillian Lathey, and Monika Wozniak, eds., *Cinderella across Cultures New Directions and Interdisciplinary Perspectives* (Detroit: Wayne State University Press, 2016), 13–31.
Henry, Astrid, *Not My Mother's Sister: Generational Conflict and Third-Wave Feminism* (Bloomington: Indiana University Press, 2004).
Hermann, Isabella, "Artificial Intelligence in Fiction: Between Narratives and Metaphors", *AI & Society* 38, (2023), 319–29.

Herndon, Cory J. and Scott McGough, *Lorwyn* (Stuttgart: Panini Books, 2007).
——, *Morningtide* (Renton, WA: Wizards of the Coast, 2008a).
——, *Eventide* (Renton, WA: Wizards of the Coast, 2008b).
Hewlett, Maurice, *John o'London* cited in Arthur Conan Doyle, *The Coming of the Fairies* (London: Hodder & Stoughton, 1922).
Hewson, David, "Cottingley Fairies a Fake, Woman says", *The Times of London* (18 March, 1983), 3.
——, "Secrets of two famous hoaxers", *The Times of London* (April 4, 1983), 3.
Hield, Fay, and Kevan Manwaring, "Special Issue on Perfoming Fairy", *Revenant* 6 (2021), 1–8.
Holden, Robert, and Nicholas Holden, *Bunyips: Australia's Folklore of Fear* (National Library of Australia, 2001).
Holmes, John, *The Pre-Raphaelites and Science* (New Haven, CT: Yale University Press, 2018).
Houlbrook, Ceri, and Simon Young, *Magical Folk: British and Irish Fairies 500AD to the Present* (London: Gibson Square, 2018).
Hughes, William, and Ruth Heholt, *Gothic Britain: Dark Places in the Provinces and Margins of the British Isles* (Cardiff: University of Wales Press, 2018).
Hussey, Christopher, *The Picturesque: Studies in a Point of View*, 3rd edn (Abingdon, Oxen: Routledge, 2018).
Ingold, Tim, "Rethinking the Animate, Re-Animating Thought", *Ethnos* 71/1 (2011), 9–20.
Irish Post, "Explore Irish Mythology: Changelings", 22 December 2021. <https://www.irishpost.com/life-style/exploring-the-irish-mythology-changelings-170347#:~:text=Changelings%20were%20fairies%20who%20had,%2C%20or%20for%20malice%2Frevenge>, accessed 11 March 2024.
Irwin, William, George, A. Dunn, and Rebecca Housel, *True Blood and Philosophy* (New Jersey: John Wiley & Sons, 2011).
Jackson, Tony E., "Imitative Identity, Imitative Art, and A.I. Artificial Intelligence", *Mosaic: An Interdisciplinary Critical Journal* 50/2 (2017), 47–63.
Jacobs, Joseph, "Childe Rowland", in *Folklore* 2/2 (June 1891), 182–97.
Jess-Cooke, Carolyn, *Film Sequels: Theory and Practice from Hollywood to Bollywood* (Edinburgh: Edinburgh University Press, 2009).
Jirsa, Curtis. R. H., "In the Shadow of the Ympe-Tre: Arboreal Folklore in Sir Orfeo" *English Studies* 89/2 (2008). 141–51.
Johanek, Peter, "König Arthur und die Plantagenets: Über den Zusammenhang von Historiographie und höfischer Epik in mitteralterlicher Propaganda", *Frühmittelalterliche Studien* 21 (1987), 346–89.
Jones, Leslie Ellen, "Fairies", in Carl Lindahl, John McNamara, and John Lindow, eds, *Medieval Folklore* (Oxford: Oxford University Press, 2002), 128–30.

Ju Lee, Young, "What Today's Children Read from "Happily Ever After" Cinderella stories", *Pedagogies: An International Journal* 17 (2022), 37–53.
Kaderabek, Kat, "Fate: The Winx Club Flop", *The Tower: Catholic University's Independent Newspaper*, 11 February 2021, <https://cuatower.com/2021/02/fate-the-winx-club-flop/>, accessed 11 March 2024.
Keene, Melanie, *Science in Wonderland: The Scientific Fairy Tales of Victorian Britain* (Oxford: Oxford University Press, 2015).
Keightley, Thomas, *The Mythology of Fairies: The Tales and Legends of Fairies from All Over the World* (London: Good Press, 2023).
Kennedy, Melanie, "Come On, Let's Go Find Your Inner Princess: (Post-)Feminist Generationalism in Tween Fairy Tales", *Feminist Media Studies* 18 (2018), 424–39.
Khair, Tabish, *The Gothic, Postcolonialism, and Otherness: Ghosts from Elsewhere* (Basingstoke: Palgrave, 2009).
King, Wilma, "'Prematurely Knowing of Evil Things': The Sexual Abuse of African American Girls and Young Women in Slavery and Freedom", *The Journal of African American History* 99/3 (2014), 173–96.
Kingsley, Charles, *The Water Babies* (Hertfordshire: Wordsworth Classics, 1994).
Kipling, Rudyard, *Puck of Pook's Hill* [1906] (London: Penguin Books, 1994).
Klapcsik, Sandor, *Liminality in Fantastic Fiction: A Poststructuralist Approach* (Jefferson, NC: McFarland & Co, 2012).
Klenk, Margaret, "Cinderella Longs for the Fairy Godmother", *Jung Journal* 17 (2023), 78–82.
Knight, Sirona, *Faery Magick: Spells, Potions, and Lore from the Earth Spirits* (Newburyport: Red Wheel/Weiser, 2002).
Koch, John T., *Celtic Culture: A Historical Encyclopedia* (New York: ABC-CLIO, 2006).
Koven, Mikel, J. (2012) "'I'm a Fairy? How Fucking Lame!: True Blood as Fairytale", in Brigid Cherry, ed., *True Blood: Investigating Vampires and Southern Gothic* (London, New York: I.B. Tauris), 59–73.
Kreider, Tim, "Review: A.I. Artificial Intelligence", *Film Quarterly* 56/2 (2002), 32–9.
Kress, Tricia, M., "'Why Do They All Have Powers?' De/Constructing Southern 'Otherness' in True Blood", *Counterpoints* 434 (2014), 107–17.
Kruse, John, *The Spirits of the Land: Faeries and the Soul of Britain* (independently published, 2022).
Kruse, John, *Faeries in the Natural World.* (Street: Green Magic, 2021).
Lackey, Mercedes, *Born to Run* (New York: Baen, 1992).
Larrington, Carolyne, *The Land of the Green Man* (London: I.B. Tauris, 2015).
——, *The Land of the Green Man* [2015] (London: Bloomsbury Academic, 2019).
Le Guin, Ursula K., "The critics, the monsters, and the fantasists", *The Wordsworth Circle* 38/1–2 (2007), 83–7.

Lecouteux, Claude, *The Tradition of Household Spirits: Ancestral Lore and Practices* (Vermont: Inner Traditions, 2000).

Lee-Ryder, Danielle, and Curtis Roman, Curtis, "Finding elusive little people". *Origins* 2 (2018), 8–9.

Legends and Traditionary Stories (London: James Burns, 1843).

Letcher, Andy, "The Scouring of the Shire: Fairies, Trolls and Pixies in Eco-Protest Culture", *Folklore* 112/2 (2001), 147–61.

Lethbridge, Stefanie, "The Wisdom of the Folly: Co-Operative Diversity in Ben Aaronovitch's Rivers of London Series", in Oliver von Knebel Doeberitz and Ralf Schneider, eds., *London Post-2010 in British Literature and Culture* (Leiden: Brill, 2017), 235–53.

Lewis, C.S., *The Discarded Image: An Introduction to Medieval and Renaissance Literature* (Cambridge: Cambridge University Press, 1964).

Libbey, Dirk, "Wait, Disneyland is replacing flying Tinkerbell in its new fireworks show?" in *Cinema Blend*, 22 December 2022. <https://www.cinemablend.com/theme-parks/wait-disneyland-is-replacing-flying-tinkerbell-in-its-new-fireworks-show#:~:text=Instead%20of%20seeing%20Tinkerbell%20fly,park%20on%20January%2027%2C%202023>, accessed 18 June 2023.

Lymbou, Paulina, "The Wicked Witch of the Discworld: A Re-examination of Magical Authority and Gender Politics", PhD dissertation (Aristotle University Thessaloniki, 2015).

Maas, Jeremy, "Victorian Fairy Painting", in L. Lambourne and J. Martineau, eds, *Victorian Fairy Painting* (Royal Academy of Arts, 1997), 10–21.

Machen, Arthur, *The White People* (1904), Project Gutenberg Australia. <http://gutenberg.net.au/ebooks06/0601371h.html>, accessed 20 July 2023.

———, *The Shining Pyramid* (1906), Project Gutenberg Australia. <http://gutenberg.net.au/ebooks06/0606971h.html>, accessed 20 July 2023.

MacRitchie, David, *The Testimony of Tradition*. (London: Kegan Paul, Trench, Trübner and Co., Ltd, 1890).

Magliocco, Sabina, "'Reconnecting to Everything': Fairies in Contemporary Paganism", in *Fairies, Demons, and Nature Spirits: "Small Gods" at the Margins of Christendom* (2018), 325–47.

———, "The Taming of the Fae: Literary and Folkloric Fairies in Modern Paganisms", in Shai Feraro and Ethan Doyle White, eds., *Magic and Witchery in the Modern West* (London: Palgrave Macmillan, 2019). 107–31.

Makoto, Hayashi and Matthias Hayek, "Onmyodo in Japanese History", *Japanese Journal of Religious Studies* 40/1 (2013), 1–18.

March-Russell, Paul, "Pagan Papers: History, Mysticism, and Edwardian Childhood", in Adrienne E. Gavin and Andrew F. Humphries, ed., *Childhood in Edwardian Fiction: Worlds Enough and Time* (New York: Palgrave Macmillan, 2009), 23–36.

Marini-Maio, Nicoletta and Ellen Nerenberg. "The 'Angelification' of Girls: *Winx Club* as a Neo-Liberal Catholic Project", *Journal of Italian Cinema & Media Studies* 8/1 (2020), 23–41.

Mason, Sophie, "Interview with Louisa John-Krol", *Feather of the Firebird*, 14 May 2018. <https://firebirdfeathers.com/2018/05/14/interview-with-louisa-john-krol>, accessed December 2023.

McCann, Michelle Roehm, and Marianne Monson-Burton, *Finding Fairies: Secrets for Attracting Little People from Around the World* (Portland: Beyond Words Publishing, 2001).

McCarron, Reilly, "The elusive Australian fairy tale", *Griffith Review* 42, 2013. < https://www.griffithreview.com/articles/8954/>, accessed 18 June 2023.

McClintock, Pamela, "How Tinker Bell became Disney's stealthy $300 million franchise", *The Hollywood Reporter*, 3 April 2014, <https://www.hollywoodreporter.com/news/general-news/how-tinker-bell-became-disneys-692559/>, accessed 18 June 2023.

McFarland Taylor, Sarah, "What If Religions Had Ecologies? The Case for Reinhabiting Religious Studies", *Journal for the Study of Religion, Nature and Culture* 1/1 (2007), 129–38.

McGrath, Thomas, "Fairy Faith and Changelings: The Burning of Bridget Cleary in 1895",*Studies: An Irish Quarterly Review* 71/282 (Summer 1982), 178–84.

McGregor, Rafe, "The Urban Zemiology of *Carnival Row*: Allegory, Racism and Revanchism", *Critical Criminology* 29 (2021), 367–83.

Mendelsohn, Farah, "Faith and Ethics", in Andrew M. Butler, Edward James, and Farah Mendelsohn, eds, *Terry Pratchett: Guilty of Literature*. Vol. 2. Foundation Studies in Science Fiction (Reading: The Science Fiction Foundation, 2001), 145–61.

Menefee, Samuel Pyeatt. "Circling as an Entrance to the Otherworld". *Folklore* 96/1 (1985), 3–20.

Meyers, Eric M., Julia P. McKnight, and Lindsey M. Krabbenhoft, "Remediating tinker bell: exploring childhood and commodification through a century-long transmedia narrative", *Jeunesse: Young People, Texts, Cultures* 6/1 (2014), 95–118.

Michaelsen, Shannen, "New Look at 'Life Size' Tinker Bell for Disney Parks, Meet and Greet May Still Debut", *Walt Disney World News Today* (10 March 2023), <https://wdwnt.com/2023/03/josh-damaro-shows-off-tiny-tinker-bell-meet-and-greet-may-still-debut/>, accessed 18 June 2023.

Mickalites, Carey, "Fairies and a Flâneur: J.M. Barrie's Commercial Figure of the Child" in *Criticism* 54 (2012), 1–27.

"Mimih", Binink Kunwok dictionary, n.d. <https://www.njamed.com/#mimih>, accessed 18 June 2023.

McAnally, David Russel. "*Irish Wonders: The Ghosts, Giants, Pookas, Demons, Leprechawns, Banshees, Fairies, Witches, Widows, and other Marvel of the Emerald Isle: Popular Tales as told by the people*. New York: Gramercy Books, 1996.

Moen, Kristian, *Film and Fairy Tales: The Birth of Modern Fantasy* (London: I.B. Tauris, 2013).

Moitra, Angana, "Fairy Genealogy in Tudor England" in Giuseppe Cusa and Thomas Dorfner eds., *Genealogisches Wissen in Mittelalter und Früher Neuzeit: Konstruktion—Darstellung—Rezeption* (Oldenbourg: De Gruyter, 2024), 361–76.

Moltenbrey, Karen, "Dark Days: Creating the Thrilling VFX of *Carnival Row*'s Flashback Episode", *Computer Graphics World* 43/2 (2020), 41–4. <https://www.sciencedirect.com/science/article/pii/S2211464522000550>, accessed 11 March 2024.

Monaghan, Patricia, ed., *The Encyclopedia of Celtic Mythology and Folklore* (New York: Facts on File, 2004).

Monteiro, João, "Interview: Caprice", in *Equilibrium Music.com*, Jan 2002. <https://www.equilibriummusic.com/interviews-0102.php>, accessed February 2024.

Moreton-Robinson, Aileen, "I Still Call Australia Home: Indigenous Belonging and Place in a White Postcolonizing Society", in Sarah Ahmed, ed., *Uprootings/Regroudings: Questions of Home and Migration* (Oxford: Berg Publishing, 2003), 23–40.

Morgan, Robin, *Sisterhood is powerful: An anthology of writings from the women's liberation movement* (New York: Vintage, 1970)

Morrissey, Thomas, "Growing nowhere: Pinocchio subverted in Spielberg's A.I. Artificial Intelligence", *Extrapolation* 45/3 (2004), 249–62.

Mukherjea, Ananya, "Mad, Bad and Delectable to Know: True Blood's Paranormal Men and Gothic Romance", in Brigid Cherry, ed., *True Blood: Investigating Vampires and Southern Gothic* (London, New York: I.B. Tauris), 109–21.

Mullis, Justin, "Cryptofiction! Science Fiction and the Rise of Cryptozoology", in Darryl Caterine and John Morehead, eds, *The Paranormal and Popular Culture: A Postmodern Religious Landscape* (New York: Routledge, 2019).

Munro, Joyce Underwood, "The Invisible Made Visible: The Fairy Changeling as a Folk Articulation of Failure to Thrive in Infants and Children", in Péter Narváez, ed., *The Good People: New Fairylore Essays* (Lexington: The University Press of Kentucky, 1997), 251–83.

Murray, Jack, "Igniting the Spark. Analog to Digital Adaptation of Narrative Affect and Player Subjectivity in Magic: The Gathering", in Phil Lopes, Filipe Luz, Antonios Liapis and Henrik Engström, eds., *Proceedings of the 18th International Conference on the Foundation of Digital Games (FDG)* (New York: ACM, 2023), 1–8.

Murray, James A.H., ed., *The Romance and Prophecies of Thomas of Erceldoune* (Early English Text Society Original Series 61 (London: N. Trübner & Co., 1875).

Mynhardt, Joe, and Eugene Johnson, *Where the Nightmares Come From: The Art of Storytelling in the Horror Genre* (Ponce de Leon: Crystal Lake, 2017).

Naremore, James, "Love and Death in *A.I. Artificial Intelligence*", *Michigan Quarterly Review* XLIV/ 2 (2005). < http://hdl.handle.net/2027/spo.act2080.0044.210>, accessed 11 March 2024.

National Folklore Schools Collection Volume 0028, p. 0232, n.d. < https://www.duchas.ie/en/cbes/volumes?Page=1&PerPage=20>, accessed 11 March 2024.

National Folklore Schools Collection Volume 0086, p. 9, n.d. < https://www.duchas.ie/en/cbes/volumes?Page=1&PerPage=20>, accessed 11 March 2024.

National Folklore Schools Collection Volume 0101, p. 125. n.d. < https://www.duchas.ie/en/cbes/volumes?Page=1&PerPage=20>, accessed 11 March 2024.

National Folklore Schools Collection Volume 0130, p. 521, n.d. < https://www.duchas.ie/en/cbes/volumes?Page=1&PerPage=20>, accessed 11 March 2024.

——— 0145, p. 149, n.d. < https://www.duchas.ie/en/cbes/volumes?Page=1&PerPage=20>, accessed 11 March 2024.

——— 0155, p. 0541, n.d. < https://www.duchas.ie/en/cbes/volumes?Page=1&PerPage=20>, accessed 11 March 2024.

——— 0168, p. 284, n.d. < https://www.duchas.ie/en/cbes/volumes?Page=1&PerPage=20>, accessed 11 March 2024.

——— 0510, p. 021, n.d. < https://www.duchas.ie/en/cbes/volumes?Page=1&PerPage=20>, accessed 11 March 2024.

——— 0546, p. 092, n.d. < https://www.duchas.ie/en/cbes/volumes?Page=1&PerPage=20>, accessed 11 March 2024.

——— 0574, p. 050–1, n.d. < https://www.duchas.ie/en/cbes/volumes?Page=1&PerPage=20>, accessed 11 March 2024.

——— 0834, p. 326, n.d. < https://www.duchas.ie/en/cbes/volumes?Page=1&PerPage=20>, accessed 11 March 2024.

——— 0956, p. 207, n.d. < https://www.duchas.ie/en/cbes/volumes?Page=1&PerPage=20>, accessed 11 March 2024.

National Science and Media Museum, "The Story of the Cottingley Fairies Shows that Image Manipulation is Nothing New", *NSMM*, 20 July 2012. <https://blog.scienceandmediamuseum.org.uk/the-story-of-the-cottingley-fairies-shows-that-image-manipulation-is-nothing-new/>, accessed 1 August 2023.

Nesbit, Edith, "The Sums That Came Right" in *Nine Unlikely Tales* (London: Ernest Benn Limited, 1901), 223–41.

———, *Wings and the Child or, the Building of Magic Cities* [1913] (London: Hodder and Stoughton, 2016).

New Daughter, The, dir. Luiso Berdejo (Anchor Bay, 2009).

Ní Dhúill, Orla, "Do Fantasy Writers Think Irish Is Discount Elvish?", *Naturally Orla* (24, August 2019), <https://naturallyorla.com/2019/08/24/do-american-writers-think-irish-is-public-domain-elvish/>, accessed 8 September 2021.

Nichols, John, *The Progresses and Public Processions of Queen Elizabeth*, 3 vols (London: Society of Antiquaries, 1823).

Niiler, Lucas P., "Green Reading: Tolkien, Leopold, and the Land Ethic", *Journal of the Fantastic in the Arts* 10/39 (1999), 276–85.

Nikolajeva, Maria, "Edith Nesbit – The Maker of Modern Fairy Tales", *Merveilles & Contes* 1 (1987), 31–44.

Noone, Kristin, "Shakespeare in Discworld: Witches, Fantasy, and Desire", *Journal of the Fantastic in the Arts* 21/1 (2010), 26–40.

O Brien, Lora, *Fairy Faith in Ireland: History, Tradition, and Modern Pagan Practice* (County Waterford: Eel and Otter Press, 2021).

Ó'Giollán, Dairmuid, "The Fairy Belief and Official Religion in Ireland", in Peter Narvaez, ed., *The Good People: New Fairylore Essays* (Lexington: University Press of Kentucky, 1991).

O'Guilin, Peadar, *The Call* (Oxford: Scholastic, 2016).

O'Regan, Marie, "Interview with John Connolly", in Joe Mynhardt and Eugene Johnson, eds., *Where the Nightmares Come From: The Art of Storytelling in the Horror Genre* (Ponce de Leon: Crystal Lake, 2017), 155–62.

Oladele, Bashirat, "Netflix's 'Fate: The Winx Saga' Whitewashes Original and Loses Its Impact", *Teen Vogue* (25 January 2021), <https://www.teenvogue.com/story/fate-the-winx-saga-whitewashes-original-loses-impact-op-ed>, accessed 11 March 2024.

Orenstein, Peggy, "What's wrong with Cinderella?" in *The New York Times* (24 December 2006), <http://popcultureandamericanchildhood.com/wp-content/uploads/2012/04/What%E2%80%99s-Wrong-With-Cinderella_-NYTimes.pdf>, accessed 18 June 2023.

Ostling, Michael, "Introduction: Where've All the Good People Gone?" *Fairies, Demons, and Nature Spirits: "Small Gods" at the Margins of Christendom* (2018), 1–53.

Outhwaite, Rentoul, Ida, *Elves and Fairies* (Melbourne: Lothian, 1916).

Owen, Alex, "'Borderland Forms': Arthur Conan Doyle, Albion's Daughters, and the Politics of the Cottingley Fairies", *History Workshop* 38 (1994), 48–85.

Pang, Carolyn, "Uncovering Shikigami: The Search for the Spirit Servant of Onmyodo", *Japanese Journal of Religious Studies* 40/1 (2013), 99–129.

Park, Sojin, "'Magic Imperialism': The Logic of Magic in Edith Nesbit's Fantasy Novels", 영어영문학 56 (2010), 501–17.

Parsons, Coleman, *Witchcraft and Demonology in Scott's Fiction* (Edinburgh: Oliver & Boyd LTD, 1964).

Paterek, Daria, "Whitewashing in Netflix's 'Fate: The Winx Saga' and How it Represents a Bigger Problem within the Industry", *Impact Magazine*, 30 January 2021. <https://

impactnottingham.com/2021/01/whitewashing-in-netflixs-fate-the-winx-saga-and-how-it-represents-a-bigger-problem-within-the-industry/>, accessed 11 March 2024.

Pedro, Dina, "Immigration, Passing, and the Racial Other in Neo-Victorian Imperialist Fiction: The Case of *Carnival Row* (2019–)", *Adaptation* 15/2 (2021), 244–63.

Pelea, Cringuta Irina, "Mirroring Cultural Fear, Anxiety and Dystopia in American Cinematography: The Movie *A.I.* (2001)", *Colloquia Humanistica* 11 (2022), 1–22.

Péporté, Pit, "Melusine and Luxembourg: A Double Memory", in Urban et al., eds, *Melusine's Footprint* (Leiden: Brill, 2017), 162–79.

Perrault, Charles, "Toads and Diamonds" in Andrew Lang, ed., *Blue Fairy Book* (Long: Longmans, Green and Co, 1889), 274–7.

Peterson, Mark Allen, "From Jinn to Genies: Intertextuality and the Making of Global Folklore", in Sharon R. Sherman and Mikel J. Koven, eds, *Folklore/Cinema: Popular Film as Vernacular Culture* (Boulder: Utah State University Press, 2007), 93–112.

Piatti-Farnell, Lorna. 2018. "Blood Flows Freely: The Horror of Classic Fairy Tales", in Kevin Corstorphine and Laura R. Kremmel, eds, *The Palgrave Handbook to Horror Literature* (Cham: Palgrave Macmillan, 2018).

Pisters, Patricia, *New Blood in Contemporary Cinema: Women Directors and the Poetics of Horror* (Edinburgh: Edinburgh University Press, 2020).

Pomerance, Murray, "Tinker Bell, the Fairy of Electricity", in Allison Kavey, ed., *Second Star to the Right: Peter Pan in the Popular Imagination* (New Brunswick: Rutgers University Press, 2009), 13–49.

Pratchett, Terry and Simpson Jacqueline, "Sir Terry Pratchett in conversation with Dr. Jacqueline Simpson", *Discworld Convention* (26 August 2010), <https://www.tandf.co.uk//journals/terrypratchett/documents/transcripts/terry-pratchett-podcast-part-two.pdf>, accessed 7 August 2021.

Pratchett, Terry, *Lords and Ladies* (London: Victor Gollanz, 1992).

——, "Imaginary Worlds, Real Stories", *Folklore* 111/2 (2000), 159–68.

——, "*Why Gandalf Never Married*", *Xyster* 11 (1986), <https://ansible.uk/misc/tpspeech.html>, accessed 7 July 2021.

Pretty Woman. dir. Garry Marshall (Touchstone Pictures, 1990).

Princess Diaries, The. dir. Garry Marshall (Walt Disney Pictures, 2001).

Purkiss, Diane, *Troublesome Things: A History of Fairies and Fairy Stories* (London: Penguin Books, 2000).

——, "Sounds of Silence. Fairies and Incest in Scottish Witchcraft Stories", in S. Clark, ed., *Languages of Witchcraft* (New York: St Martin's Press, 2001).

——, *Fairies and Fairy Stories: A History* (Stroud: Tempus, 2007).

Putter, Ad, "Finding Time for Romance: Mediaeval Arthurian Literary History", *Medium Aevum* 63/1 (1994), 1–16.

Radner, Hilary, *Neo-feminist Cinema: Girly Films, Chick Flicks and Consumer Culture* (New York: Routledge, 2011).

Recchio, Devin T., "Constructing Abe no Seimei: Integrating Genre and Desparate Narratives in Yumemakura Baku's *Onmyoji*", MA thesis (University of Massachusetts, Amherst, 2014), <https://scholarworks.umass.edu/cgi/viewcontent.cgi?article=1134&context=masters_theses_2.>, accessed 11 March 2024.

Reider, Noriko T., "Onmyoji: Sex, Pathos, and Grotesquery in Yumemakura Baku's Oni", *Asian Folklore Studies* 66/1-2 (2007), 107-24.

Renner, Karen J, "Changelings", in Karen J. Renner, ed., *Evil Children in the Popular Imagination* (London: Palgrave Macmillan, 2016), 153-75.

Rikhardsdottir, Sif, "Chronology, Anachronism and *Translatio Imperii*", in Leah Tether and Johnny McFadyen, eds., *Handbook of Arthurian Romance: King Arthur's Court in Medieval European Literature* (Berlin: De Gruyter, 2017), 135-50.

Robbins, Joel, "Crypto-Religion and the Study of Cultural Mixtures: Anthropology, Value, and the Nature of Syncretism", *Journal of the American Academy of Religion*, 79/2 (2011), 408-24.

Roman, Curtis, "Indigenous Beliefs About Little People", *ab-Original* 3/1 (2019), 124-29.

Romano, Renee Christine, *Race Mixing: Black–White Marriage in Postwar America* (Cambridge, MA: Harvard University Press, 2009).

Rone, Vincent, "Scoring the Familiar and Unfamiliar in Howard Shore's 'The Lord of the Rings'" *Music and the Moving Image* 11/2 (2018), 37-66.

Rothwell, Nicolas, "In Search of the Indigenous Little People of Northern Australia", *The Australian* (2014), <https://www.theaustralian.com.au/arts/review/in-search-of-the-indigenous-little-people-of-northern-australia/news-story/4b6a79f0661a7edoc55ae00e29fb119a>, accessed 11 March 2024.

Rozario, Do. Rebecca-Anne, "Australia's Fairy Tales Illustrated in Print Instances of Indigeneity, Colonization, and Suburbanization", *Marvels & Tales* 25/1 (2011), 13 –32.

Rummell, Kathryn, "Rewriting the Passing Novel: Danzy Senna's Caucasia", *The Griot*, 26/2 (2007), 1-14.

Said, Edward, *Orientalism* (London: Routledge, 1978).

Saler, Michael, "'Clap If You Believe in Sherlock Holmes': Mass Culture and the Re-enchantment of Modernity, c. 1890-c.1940", *The Historical Journal* 46 (2003), 599-622.

Sasani, Samira, "The Colonized (the Other) and the Colonizer's Response to the Colonial Desire of 'Becoming Almost the Same But Not Quite the Same' in *M. Butterfly*", *Journal of Language Teaching and Research*, 6/2 (2015), 435-42.

Saunders, Corinne, *Magic and the Supernatural in Medieval English Romance* (Cambridge: D. S. Brewer, 2010).

Sawers, Naarah, "Building the Perfect Product: The Commodification of Childhood in Contemporary Fairy Tale Film", in Sidney Eve Matrix, ed., *Fairy Tale Films: Visions of Ambiguity* (Logan: Utah State University Press, 2010), 42–9.

Scheible, Ellen, "The Danger of the Domestic in Ireland: Bridget Clear, Big HouseModernism, and Tana French", *Tulsa Studies in Women's Literature* 41/1 (2020), 113–33.

Seal, Graham, "Primal Evil" *Grisly History* (13 February 2018), <https://gristlyhistory.blog/tag/long-lankin/>, accessed 1 August 2021.

Sex and the City, created by Darren Starr (Home Box Office, 1998–2004).

Shanahan, J., *Terry Pratchett:* "Mostly Human", in Bernice M. Murphy, ed., *Twenty-First-Century Popular Fiction* (Edinburgh: Edinburgh University Press, 2018), 31–40.

Shinichi, Shigeta, "A Portrait of Abe no Seimei", *Japanese Journal of Religious Studies* 40/1 (2013), 77–97.

Shippey, Tom, "The Faërie World of Michael Swanwick", in Dimitra Fimi and Thomas Honegger, eds., *Sub-Creating Arda: World-Building in J.R.R. Tolkien's Works, Its Precursors and Its Legacies* (Zollikofen: Walking Tree Publishers, 2019), 415–29.

Sikora, Joshua, "The Everlasting Moment: Enchantment and Myth in *A.I.* and *2001: A Space Odyssey*", in Elisa Colombani, ed., *A Critical Companion to Stanley Kubrick* (London: Lexington Books, 2020), 263–77.

Silver, Carole, "On the Origin of Fairies: Victorians, Romantics, and Folk Belief", in *Browning Institute Studies*, Vol. 14, The Victorian Threshold (1986), 141–56.

———, *Strange and Secret Peoples: Fairies and Victorian Consciousness* (Oxford: Oxford University Press, 2000).

———, "Faerie and Fairy Lore", in Donald Haase, ed., *The Greenwood Encyclopedia of Folktales and Fairy Tales* (Westport, CT: Greenwood Press, 2008a), 318–320.

———, "Fairy, Fairies", in Donald Haase, ed., *The Greenwood Encyclopedia of Folktales and Fairy Tales* (Westport, CT: Greenwood Press, 2008b), 321–2.

Simpson, Jacqueline, "The Folklore of Infant Deaths: Burials, Ghosts and Changelings", in Gillian Avery and Kimberley Reynolds, eds, *Representations of Childhood Death* (London: Palgrave Macmillan, 2000), 11–28.

———, "On the Ambiguity of Elves", *Folklore* 122/1 (2011), 76–83.

Simpson, Jacqueline, and Steve Roud, "Changelings", in *A Dictionary of English Folklore* (Oxford: Oxford University Press, 2000a), 53–4.

———, "Fairies", in *A Dictionary of English Folklore* (Oxford: Oxford University Press, 2000b), 115–16.

———, "Fairyland", in *A Dictionary of English Folklore* (Oxford: Oxford University Press, 2000c), 116–17.

———, "Goblin", in *A Dictionary of English Folklore* (Oxford: Oxford University Press, 2000d), 146.

Sinclair, Lian, "Magical Genders: The Gender(s) of Witches in the Historical Imagination of Terry Pratchett's Discworld", *Mythlore: A Journal of J.R.R. Tolkien, C.S. Lewis, Charles Williams, and Mythopoeic Literature* 33/2, article 4 (2015).

Smith, Barbara, "The Expression of Social Values in the Writing of E. Nesbit" *Children's Literature* 3 (1974), 153–64.

Smith, Michelle, "How Early Australian Fairy Tales Displaced Aboriginal People with Mythical Creatures and Fantasies of Empty Land", *The Conversation* (6 July 2022), <https://theconversation.com/how-early-australian-fairy-tales-displaced-aboriginal-people-with-mythical-creatures-and-fantasies-of-empty-land-185592>, accessed 11 March 2024.

Smith, Paul. "The Cottingley Fairies: The End of a Legend", in Peter Narvaez, ed., *The Good People: New Fairylore Essays* (London: Garland Publishing Inc, 1991), 371–405.

Spires, Ashley, *Fairy Science* (New York: Crown Books, 2019).

Stafford, Cooke, "Fairy Mythology", in Thomas Hood, ed., *Hood's Magazine and Comic Miscellany,* Vol. X (July–December, 1848), 330–55.

Stasiewicz-Bieńkowska, Agnieszka, "Lustful Ladies, She-Demons and Good Little Girls: Female Agency and Desire in the Universes of Sookie Stackhouse", Continuum 33/2 (Taylor & Francis Online, 2019), 230–41.

States News Service, "Disney's 'Tinker Bell and Lost Treasure' to premiere at United Nations Headquarters, with Tinker Bell to be Named 'Honorary Ambassador of Green'", *States News Service,*22 October 2009.

Strodder, Chris, *Disneyland Encyclopedia: The Unofficial, Unauthorized, and Unprecedented History of Every Land, Attraction, Restaurant, Shop, and Event in the Original Magic Kingdom* (Santa Monica: Santa Monica Press, 2008).

Sturgis, Amy H., "'Tolkien is the Wind and the Way': The Educational Value of Tolkien-Inspired World Music", in Bradford D. Eden, ed., *Middle Earth Minstrel: Essays on Music in Tolkien, Jefferson, McFarland & Company* (2010), 126–39.

Sudha, S., and S. Irudaya Rajan, "Female Demographic Disadvantage in India 1981–1991: Sex Selective Abortions and Female Infanticide," *Development and Change* 30 (1999), 585–618.

Sugg, Richard, *Fairies: A Dangerous History* (London: Reaktion Books, 2018).

Summers, David A., *Spenser's Arthur: The British Arthurian Tradition and The Faerie Queene* (Lanham and Oxford: University Press of America, 1997).

Sundmark, Björn, "Andrew Lang and the Colour Fairy Books," *IRSCL Congress* (Dublin, Ireland, 2005).

Swanwick, Michael, *The Iron Dragon's Daughter* [1994] (SFBC, 2007).

Tatar, Maria, "Female Tricksters as Double Agents", The Cambridge Companion to Fairy Tales (Cambridge: Cambridge University Press, 2014a), 39–59.

———, "Introduction", in Maria Tatar, ed., *The Cambridge Companion to Fairy Tales* (Cambridge: Cambridge University Press, 2014b), 1–10.
Terrible Tragedy in Kerry. A Family of Lunatics. A Mother Murders Her Idiot Son", *The Irish Examiner*, 1 February 1888.
"The Tragedy Near Killarney", *Kerry Sentinel*, 3 February 1888.
Thompson, Stith, *Motif Index of Folk Literature* (Copenhagen: Rosenkilde and Bogger, 1958).
Thorne, William, "The Folk-Lore of Shakespeare", *The Athenaeum. Journal of Literature, Science and the Fine Arts* (1847), 981–3.
Tibbetts, John, "Robots Redux A.I. Artificial Intelligence", *Literature/Film Quarterly* 2 9/4 (2001), 256–61.
Tolkien, J. R. R., "On Fairy-Stories", in Verlyn Flieger and Douglas Anderson, eds., *Tolkien On Fairy-Stories* (London: HarperCollins, 2014), 27–84.
Tralee, Saturday, "July 16, 1887", *Kerry Evening Post*, 16 July 16 1887.
Trosper, R. L., *Indigenous Economics: Sustaining Peoples and Their Lands* (Phoenix, AZ: University of Arizona Press, 2022).
True Blood, Seadon 1, Episode 2, "Strange Love", dir. Alan Ball (HBO, 7 September 2008).
———, Season 3, Episode 12, "Evil is Going On", dir. Anthony Hemingway (HBO, 12 September 2010).
———, Season 3, Episode 9, "Everything is Broken", dir. Scott Winant (HBO, 15 August 2010).
———, Season 4, Episode 1, "She's Not There", dir. Michael Lehmann (HBO, 26 June 2011).
U.S. Department of Health and Human Services, "Batten Disease and Other Neuronal CeroidLipofuscinoses" (July 2018).
Van Muylem, Filip, "Daemonia Nymphe", *Peek-a-boo Music Magazine* (16 January 2013), <http://www.peek-a-boo-magazine.be/en/interviews/daemonia-nymphe/>, accessed December 2023.
Varner, Gary R., *Creatures of the Mist: Little People, Wild Men and Spirit Beings Around the World* (New York: Algora Publishing, 2007).
Vyse, Stuart, "The Enduring Legend of the Changeling", *Skeptical Inquirer* 42/4 (1 March 2018), <https://skepticalinquirer.org/exclusive/the-enduring-legend-of-the-changeling/>, accessed 11 March 2024.
Wade, James, *Fairies in Medieval Romance* (New York: Palgrave Macmillan, 2011).
Watkins, Carl, *History and the Supernatural in Medieval England* (New York: Cambridge University Press, 2007).
Waugh, Patrica, *Metafiction: The Theory and Practice of Self-Conscious Fiction* [1984] (London: Routledge, 2001).
Wehnert, Kathleen, *Passing an Exploration of African-Americans on Their Journey for an Identity along the Colour Line* (London: Verlag, 2010).

Weinstock, Jeffrey Andrew, ed., *The Ashgate Encyclopedia of Literary and Cinematic Monsters* (New York: Routledge, 2016).
White, Wolf, *Changeling: The Dreaming*, 2nd edn (Clarkston: White Wolf, 1997).
——, *Changeling: The Lost: A Story-Telling Game of Beautiful Madness* (Stone Mountain: White Wolf, 2007a).
——, *Winter Masques* (Stone Mountain: White Wolf, 2007b).
White, Alan, V., "*A.I. Artificial Intelligence:* Artistic Indulgence or Advanced Inquiry?", in Dean A. Kowalski, ed., *Steven Spielberg and Philosophy: We're Gonna Need a Bigger Book* (Kentucky: University of Kentucky Press, 2008), 210–26.
White, Carolyn. *A History of Irish Fairies* (New York: Hachette Books, 2005).
White, Tom, "Medieval Trees", *Dandelion* 4/2 (Winter 2013).
Wilde, Lady Francesca Speranza, *Ancient Legends, Mystic Charms and Superstitions of Ireland v.1.* (London: Global Grey Ebooks, 1887).
William of Newburgh (*Guilielmum Nevbrigensis*). *Historia Anglicana, sive de regno et administrarione Regum Angliæ* (Paris: Joannem Petit-Pas, 1882).
Williams, Tad, *The War of the Flowers* (London: Orbit, 2003).
Wilson, Jean, *Entertainments for Elizabeth I* (Woodbridge: D.S. Brewer, 1980).
Winx Club, dir. Iginio Straffi (RAI, 2004–2009).
Wolf, Werner, "Is There A Metareferential Turn, and If So, How Can it Be Explained?" in Werner Wolf, ed., *The Metareferential Turn in Contemporary Arts and Media: Forms, Functions, Attempts at Explanation* (Amsterdam: Brill, 2011), 1–47.
Wolfrum, Sophie, ed., *Porous City: From Metaphor to Urban Agenda* (Basel: Birkhauser, 2024).
Wood, Juliette, "Filming Fairies: Popular Film, Audience Response and Meaning in Contemporary Fairy Lore", *Folklore* 117/3 (2006), 279–96.
Woodcock, Matthew, *Fairy in The Faerie Queene: Renaissance Elf-Fashioning and Elizabethan Myth-Making* (Aldershot: Ashgate, 2004).
Wynne, Catherine, "Arthur Conan Doyle and Psychic Photographs", *History of Photography* 22/4 (1998), 385–92.
Yeats, William Butler, *The Poetical Works of William B. Yeats: Volume 1* (New York: The Macmillan Company, 1908).
——, *Writings on Irish Folklore, Legend and Myth*, ed. Robert Welch (London: Penguin, 1993).
Young, Diana, *FernGully* (Sydney: Ashton Scholastic, 1991).
Young, Kay "Change Your Shoes, Change Your Life: On Object Play and Transformation in a Woman's Story", *American Journal of Play* 4 (2011), 285–309.
Young, Simon, "When Did Fairies Get Wings?" in Darryl Caterine and John Morehead, eds., *The Paranormal and Popular Culture: A Postmodern Religious Landscape* (New York: Routledge, 2019).

Young, Simon, and Ceri Houlbrook, *Magical Folk: A History of Real Fairies, 500AD to the Present* (London: Gibson Square Books, 2022).

Yumemakura, Baku, *Onmyoji,* Moro Miya, trans. (Taipei: Muses Publishing House, 2003).

Zalka, Virág Csenge, "Adventures in the Classroom Creating Role-Playing Games Based on Traditional Stories for the High School Curriculumon Traditional Stories for the High School Curriculum", East Tennessee State University, *Electronic Theses and Dissertations* (2012), Paper 1469, <https://dc.etsu.edu/etd/1469>, accessed 11 March 2024.

Zanescu, Andrei, "Designing Magic: The Gathering's Amonkhet: Egyptianness and the Limitations of Cultural Resonance", in Kostas Karpouzis, Stefano Gualeni, Johanna Pirker, and Allan Fowler, eds., *Proceedings of the 17th International Conference on the Foundations of Digital Games (FDG)* (New York: ACM, 2022), 1–12.

Notes on Contributors

KRISTIN AUBEL studied Applied Literary and Cultural Studies at TU Dortmund University. She is a PhD candidate and university assistant at the Department of English and American Studies at the University of Vienna, Austria. Her research interests include myth, identity construction, migration literature, magical realism, fantasy, and superhero comics. Her PhD thesis focuses on transcultural transformations of myth in contemporary fantastic literature.

SIMON BACON is an independent scholar based in Poznań, Poland. He has written and edited over thirty books on various subjects, including *Gothic: A Reader* (2018), *Horror: A Companion* (2019), *Eco-Vampires* (2020), *Nosferatu in the 21st Century* (2023), *Future Folk Horror* (2023), *The Palgrave Handbook of the Vampire* (2024), and *The Palgrave Handbook of the Zombie* (forthcoming). He also runs the book series *Vampire Studies: New Perspectives on the Undead* at <https://www.peterlang.com/series/vsu>.

FRANCESCA BIHET is an independent scholar who completed her PhD, "Folklore and Fairies: the History of Fairies in the Folklore Society from 1878 to 1945", at the University of Chichester in 2020. Among other articles, she has published the chapters "Pouques and the Faiteaux: the Channel Islands", in *Magical Folk* (2018), and "Death and the Fairy: Hidden Gardens and the Haunting of Childhood", in *Uncanny Ecogothic Gardens in the Long Nineteenth Century* (2020).

SAGA BOKNE is a PhD student in English literature at Karlstad University, Sweden. Her research interests centre on various aspects of the fantasy genre, including the politics of fantasy, fantasy's engagement with modernity, and, most of all, the various ways in which the fantasy genre builds on, appropriates from, and reimagines folklore. She writes her doctoral dissertation on the functions and meanings of the fairy in contemporary fantasy literature.

SHANE BRODERICK is an Irish folklorist who has a wide interest in many aspects of Irish tradition and history. He has a BA (Joint Hons) in Irish Folklore and Celtic Civilisation and is currently a candidate for an MA in Irish Folklore and Mythology at University College Cork. He was recently published in *Folklore, People, and Places: International Perspectives on Tourism and Tradition in Storied Places* by Routledge.

JO ANNA BURN is an independent scholar based in New Zealand. Her ecclectic research interests include science fiction and fantasy, reality television, refugee education, legal language and legal interpreting.

MORGAN DAIMLER is an author and independent researcher focusing primarily on Irish mythology and fairylore. She has written the book *A New Dictionary of Fairies* and has presented the paper "Evolution of the Scottish Fairy Courts: from ballads to urban fantasy" at Ohio State University's "Fairies and the Fantastic" conference as well as the paper "Unseely to Antihero: the evolution of dangerous fairies in folklore, fiction, and popular belief" at Hertfordshire University's "Ill Met By Moonlight" conference.

MUSKAN DHANDHI teaches English at National Institute of Technology (NIT) Uttarakhand. She has a PhD in English Literature and Translation Studies from Indian Institute of Technology (IIT) Mandi, India. Her research interests are translation studies, folklore, and cultural studies. She was awarded the Charles Wallace India Trust Research Grant 2023 and the Shastri-Indo-Canadian MITACS Globalink Research Award 2023 for her research visits to the United Kingdom and Canada respectively. She also worked as a Research Associate in a research project pertaining to Oral History funded under IMPRESS-ICSSR (Indian Council for Social Science and Research).

ABIGAYLE FARRIER is an English Lecturer at the University of North Texas, where she teaches courses on writing and literature. Her research interests include 19th-century transatlantic women's writing, women's literary networks, 19th-century literary production, and the intersections of psychology, pedagogy, and literature. She is deeply invested in archival recovery work and has published on the first Antiguan novel, women's mental health in 19th-century literature, and feminist pedagogy.

GEMMA FILES is a Canadian horror writer, journalist, and film critic. Her short story, "The Emperor's Old Bones," won the International Horror Guild Award for Best Short Story of 1999. Five of her short stories were adapted for the television series *The Hunger*.

NICK FREEMAN is Reader in English at Loughborough University. He has published widely on the supernatural and occult, with articles and book chapters on topics such as ghost stories, modern paganism, witchcraft, and black magicians. He has also written on many Gothic, Weird, and fantasy writers, notably Arthur Machen, Vernon Lee, M. John Harrison, and Robert Aickman. He has yet to find any fairies at the bottom of his garden, but he lives in hope.

MARIA GIAKANIKI is an independent scholar and editor-in-chief of Ars Nocturna, a small publishing house in Athens, Greece. She has edited Uncanny Ireland: Otherworldly Tales of the Strange & Sublime (The British Library, 2024), and co-edited Bending to Earth: Strange Stories by Irish Women (Swan River Press, 2019). She has contributed chapters to collections of eassays published by Manchester University Press, Palgrave Macmillan, McFarland and Peter Lang. She is also a writer of uncanny and surreal micro fiction.

MARGARYTA GOLOVCHENKO is a PhD candidate in the art history department at the University of Oregon, which is situated on the land of the Kalapuya people. Her research, which is supported by the Social Sciences and Humanities Research Council of Canada, examines women-animal relationships in French and British art and visual culture of the 18th and 19th centuries. Her scholarship has appeared in publications such as Journal of Comparative Literature and Aesthetics, Fantastika Journal, Symbolism: An International Annual of Critical Aesthetics, Journal of Posthumanism, and Capacious, with book chapters forthcoming in the edited volumes Environmental and Cultural Destruction at Imperial Margins (DeGruyter) and Going Feral: A Proposition for a Speculative Animism in the Arts (Vernon Press). She is also an art and literary critic, as well as an associate editor for *Material Culture Review*.

AMY HARRIS is a fully funded doctoral researcher at De Montfort University. Her thesis considers contemporary British horror films made by women. Amy has published chapters in *Bloody Women: Women Directors of Horror*; *Evil Women: Representations within Literature, Culture and Film* (2022);" and *Folk Horror on Film: Return of the British Repressed* (2023). In her spare time, Amy posts about her PhD experience on Instagram under the handle @the_horrorhag.

LESLEY HAWKES is Associate Professor in the School of Communication, Creative Industries Faculty, Queensland University of Technology, Brisbane, Australia. Her specialist areas of research are how Australian places of belonging are created through works of fiction, spatial understanding as it relates to Australian literature and environmental studies and Australian literature.

CLAY FRANKLIN JOHNSON is the author of *A Ride Through Faerie & Other Poems* (2021), an illustrated collection of poetry published by Gothic Keats Press. His collection's eponymous poem was presented at "Ill met by moonlight": Gothic encounters with enchantment and the Faerie realms in literature and culture, a conference organised by the Open Graves, Open Minds Project (OGOM) with the University of Hertfordshire. In December 2024, Clay's poem "The Faery Wood" won the Highly Commended Award, one of two prizes given for the Brian Nisbet Poetry Award in Huntly, Scotland. His writing has been nominated for the Pushcart Prize, Rhysling Award, Elgin Award, received Honorable Mention in The Best Horror of the Year, and has appeared in publications such as *Nightingale & Sparrow*, *The Fairy Tale Magazine*, *Abyss & Apex*, and *Gramarye*, among others.

NANCY JOHNSON-HUNT is a former advertising strategist turned doctoral candidate at Auckland University of Technology. After a decade-long career within the advertising and marketing industry, both in New Zealand and North America, she returns to AUT, undertaking her PhD within the

Popular Culture Research Centre. Her doctoral thesis explores the representation of ethnicity in popular reality television dating shows and how these portrayals conform to, or challenge stereotypes historically constructed to amplify racialised bodies, spaces, and issues. Her research interests include the diffusion of advertising culture, construction of ethnic and racial identities across popular media and culture, and the influence of celebrity in shaping everyday lives.

SOPHIA LANGE is a research assistant at the chair of British Literary and Cultural Studies at TU Dortmund University, Germany. She holds an MA in English and German Studies, has received the valedictorian award for her MA Thesis on "The Function of Fate in British Fantasy," and has recently defended her PhD thesis on "Inheritance and Indebtedness in the Godwinian Novel". Her research interests include the Gothic, monsters and the monstrous, trauma studies, Cultural Studies, British Romanticism, and fantastic literature.

AMY LEE is a professor in the School of Arts and Social Sciences at the Hong Kong Metropolitan University. She has published creative non-fiction as well as critical studies in the area of contemporary feminist fiction, autobiographical writing, witchcraft and magic, and using literature for creative learning experiences. A core research and pedagogical interest of hers is deploying literary texts to enhance personal well-being. Her latest research project uses Playback Theatre to build creativity and self-understanding in inclusive communities.

J. S. MACKLEY has taught English and creative writing at the University of Northampton and British fantasy Literature at Richmond University in London. His PhD focused on the Latin and Anglo-Norman versions of the Voyage of St Brendan. He has published on English folklore and mythology and medieval, gothic, and fantasy literature, with subjects including: gothic novelist Eleanor Sleath; the Victorian anti-hero Spring-Heeled Jack; August Derleth, the author of macabre Lovecraftian tales; and the legendary television series *Doctor Who* and *Quatermass*. He has recently completed the seven-volume *Spring-Heeled Jack Library* and is currently working on a monograph analysing aspects of English mythology.

PÉTER KRISTÓF MAKAI recently finished his Crafoord Postdoctoral Fellowship in Intermedial and Multimodal Studies at Linnaeus University in Växjö, Sweden. He is set to join the University of Duisburg-Essen's Cultural Studies Institute as an International Visiting Fellow to study how theme parks are transmediated into digital and board games. He got his English Literature PhD from the University of Szeged. He has published work on Tolkien, games and worldbuilding in Reconstructing Arda, Tolkien Studies, and in Postmodern Reinterpretations of Fairies Tales. He is a member of COST Action 18230, Interactive Narrative Design for Complexity Representations.

LESLEY MCLEAN is a lecturer in philosophy and religion within the School of Humanities, Arts and Social Sciences at the University of New England. She has published in the area of animal ethics – her philosophical interest – and is currently researching the intersection of "cults" with tourism her religious studies interests.

KIRSTIN A. MILLS is Senior Lecturer and Director of the Master of Research in the Faculty of Arts at Macquarie University, Sydney, Australia. Her research specialises in dreams, the sciences of the mind, and the supernatural in Gothic and fantastic literature of the 19th century, and its adaptations into 21st-century digital and visual media. A long-time lover of fairies and all things magical, she is also an artist, and frequently paints the bewitching interactions between fairies and the natural world.

ANGANA MOITRA is Assistant Professor at O.P. Jindal Global University, India. She received her PhD on the Erasmus Mundus TEEME (Text and Event in Early Modern Europe) programme, a joint-doctoral programme funded by the European Union where she was jointly based at the University of Kent and Freie Universität Berlin. Her doctoral thesis charted the evolution and transformation of the figure of the Fairy King between the Middle Ages and the early modern period, and she is currently developing her thesis into a monograph for Palgrave Macmillan. Her academic interests range from Chaucer, Malory, and the medieval romance to Renaissance humanism, metaphysical poetry, Shakespeare, and Milton.

Notes on Contributors

JOAN ORMROD is Senior Lecturer of Film and Media Studies at Manchester Metropolitan University and the editor of Routledge's *Journal of Graphic Novels and Comics*. Her research is in popular culture particularly comics, gender, fantasy, and science fiction. Her latest publications are *Wonder Woman, the Female Body and Culture* (Bloomsbury, 2020), *Time Travel in Popular Media: Essays on Film, Television, Literature and Video Games* (McFarland, 2015).

FERNANDO GABRIEL PAGNONI BERNS is a professor at the Universidad de Buenos Aires (UBA) and teaches courses on international horror film. He is Director of the research group on horror cinema "Grite", and is part of the editorial board of New Cinemas: Journal of Contemporary Film (Intellect). He wrote *Alegorías televisivas del franquismo. Narciso Ibáñez Serrador y las historias para no dormir (1966–1982)* (2019), co-edited *The Cinema of James Wan: Critical Essays* (2021), *Horror and Philosophy: Essays on Their Intersection in Film, Television and Literature* (2023), and *Bloodstained Narratives: The Giallo Film in Italy and Abroad*, and edited *A Critical Companion to Wes Craven* (2023). He is also Director of "Terror: Estudios Críticos" (Universidad de Cádiz), the first-ever horror studies series in Spain.

KELLY PALMER researches lived experience and representations of Southeast Queensland. Her practice-led PhD thesis on the Gold Coast presents a collage of myths upon everyday life, and she is editing an upcoming book on the children's animation, *Bluey*, set in Brisbane, Australia. Kelly teaches media and communication, writing and creativity, across Queensland universities, libraries, and schools.

LORNA PIATTI-FARNELL, PhD, is Academic Dean at SAE Creative Media Institute in Auckland, New Zealand. She is the Director of the Australasian Horror Studies Network (AHSN) and an Adjunct Research Professor at Curtin University, Australia. Her research sits primarily in screen media and cinematic cultures, with a particular focus on transmedia storytelling, eco-narratives, digital technologies, popular iconographies, and a long-standing interest in Gothic horror and fantasy. Prof. Piatti-Farnell is sole editor of the Routledge Advances in Popular Culture Studies book series, and co-editor of the Horror Studies book series for Lexington/Bloomsbury.

CATALINA MILLÁN SCHEIDING, PhD, is an associate professor in the Liberal Arts Department at Berklee College of Music Valencia Campus. She is a member of the Grupo de Investigación TALIS of the University of València in language education and interculturality, and is part of a project developed by the Faculty of Teacher Training on H5P integration for the education of Sustainable Development Goals. She is also the secretary of the Journal of Literary Education, and has published in numerous international journals. Dr Millán-Scheiding's areas of research include children's poetry, multimodality, and translation, as well as narratology in fantasy and fiction.

SUMAN SIGROHA, PhD, is a researcher and teacher at Indian Institute of Technology Mandi, Himachal Pradesh, India. With her training in the fields of literary studies and psychology, she engages with texts through psycho-social concepts like stereotyping, implicit bias, memory and representation. Her recent research focuses on contemporary literature from troubled regions of India, South Asia, and America, rich with unsettling questions about nationalism, belonging, and identity. She has recently contributed to and co-edited *Translational Research and Applied Psychology in India* (SAGE, 2019).

BLAIR SPEAKMAN is a popular culture scholar who is particularly fascinated with the representation of folklore and queerness in contemporary Gothic-Horror television series. This interest led Speakman to start his PhD at Auckland University of Technology (AUT) in July 2017, focusing on Queer characters in contemporary Gothic television shows. He is highly involved in extra-curricular activities, including being a member of the Popular Culture Research Centre as well as a committee member for Out@AUT, the University's student LGBTIQA club.

ALLYSON WIERENGA is a fourth-year PhD candidate in English. Her research focuses on child agency in children's literature and intersections between children's literature, health humanities, and disability studies.

JENNY WISE is Associate Professor in Criminology in the School of Humanities, Arts and Social Sciences at the University of New England. Her research focuses upon the social impacts of forensic science on the criminal justice system, the role of the CSI Effect changing criminal justice practices, dark tourism, and crime as a form of leisure and popular culture.

REBECCA WYNNE-WALSH completed her PhD, "Basque Gothic Cinema (1990–2020): A Regionalist Challenge to the Spanish Model of National Cinema Production and Cultural Identity", with Dr Xavier Aldana Reyes at Manchester Metropolitan University. She is now a lecturer in Film Studies and Production at Edge Hill University. She received her MPhil in International History from Trinity College Dublin where she previously received her BA in Film Studies and English Literature. Her research interests include the following: film studies, cinema history, horror cinema, Gothic studies, cultural studies, Hispanic studies, regionalism, transnationalism, postcolonialism, trauma studies, and folklore studies.

Index

abduct 28, 42, 77–8, 138, 218, 246–8.
abuse 28, 33, 43, 116, 157, 209, 218, 277.
agency 8–9, 59, 103–4, 106–9, 122, 126, 169, 179, 207, 209, 214, 215–6, 265, 267, 273.
alien 9, 46.
alienation 81, 117, 220.
ancient 42, 64, 67–8, 70, 135, 139–40, 164, 169, 177, 182–3, 186, 211, 265–6.
angel 244.
— fallen 168.
angelic 11, 202.
archetype 47, 48, 104, 193, 198, 200–1.
arithmetic 10, 207, 210–4.
Arthurian 3, 56–7, 59–60, 67, 70, 228
authentic 18, 20–1, 23, 43–4, 186.
— in 21, 156.

baby 77–8, 84, 96, 113, 115–8, 137–8.
banshee 135, 253.
belief 2, 4, 5, 8–9, 11, 17–8, 22–3, 26–8, 39, 43, 45–6, 48–50, 65, 78, 82, 111–7, 136, 140, 155, 168–70, 182, 223, 233, 251, 257–60, 267–8.
birth 45, 56, 106, 112, 114, 117–8, 126, 176, 177, 256, 263.
black magic 41.
Blackness 84–5, 88, 127–9.
blue fairy 7, 13, 46–51, 273.
boggarts 9, 143–5.
Britain 30, 64–6, 71, 233–4, 243–4.
brownies 143–5, 148, 151.

Celtic 3, 67, 136, 143, 217, 223, 243, 262, 269, 271–2, 274.

changeling 7–8, 27–35, 42, 46, 68, 77–8, 111–4, 118, 136, 155, 158, 216–21.
child 25, 27–8, 31–5, 46–7, 50–1, 111–7, 155, 176, 207, 213–4, 218, 220.
child-eating 164.
childhood 28, 75–8, 114, 207–10.
childish 41, 46, 94.
children 3–5, 20–2, 68, 74–8, 94–5, 98–9, 101, 126, 128, 136, 143, 160, 167, 173, 177, 190, 194, 223, 233, 243, 245–7, 251–2, 254, 256, 263, 265, 267.
Christianity 33, 59, 74, 112, 210, 230, 244, 246, 248, 250.
Cinderella 5, 8, 47, 121–9.
city 48–9, 59, 64, 79–81, 121, 154, 158–60, 195, 197, 224.
class 4, 9, 20, 58, 121–3, 125–7, 129, 138, 145, 147, 149, 153, 159, 208–9, 217–9.
classic 3, 5, 38, 98, 122, 155, 184, 186.
Classical 7, 9, 59, 64, 195, 227, 230, 246–50.
—Neo 223–4, 228–9.
Clinkerbelle 136.
colonialism 80, 84, 176, 219, 258, 273–4, 276.
— post 80, 269, 273.
community 39, 85, 111–7, 139, 220, 275.
Conan-Doyle, Arthur 7, 17–25.
covet 77.
cruelty 46, 78, 122, 137, 139, 146, 149, 155, 219, 254.
cryptozoology 257.

dance 18–9, 63, 69, 171, 177, 188–9, 219, 223, 230, 235, 244, 248, 273.

darkness 215–6, 238, 264.
death 22–3, 27–8, 30, 32–3, 48, 64, 67–70, 78, 111–4, 117, 122, 140, 163, 170–2, 184, 217, 248, 252, 272.
deception 20, 68, 106.
demon 31, 77, 143–4, 150, 164, 230, 243–4, 250.
desire 5–6, 48, 78, 94, 96, 99, 104, 106–7, 112, 123, 127, 147, 149–50, 178, 193, 200, 211, 215, 221, 254, 265, 267.
devil 164, 168, 244–5, 249–50.
difference 8, 79–88, 108, 190.
disability 28, 31, 113.
discrimination 43, 86, 219.
Disney 5, 8, 40, 46–7, 78, 93–102, 136, 194.
diversity 10, 121, 123, 127.
divine 47, 49, 164.
doll 75, 78, 97–8.
dream 7, 9, 45, 49–51, 72, 121, 125–6, 129, 137, 147, 164, 177–9, 213, 215–6, 224, 227, 233, 248, 261–4, 267, 276.
dwarves 135, 144, 219, 272.
dystopia 42, 45, 72, 267.

eco-fairies 11.
ecology 8, 10, 79–88, 231, 264–7.
economics 105, 139, 156, 176, 268.
education 44, 58, 87, 209–10, 211–2, 214, 233–4, 237–8, 262.
Edwardian 41, 152, 162, 207–9.
electricity 194, 234–5, 238, 240, 270.
elfshot 3, 173.
elves 8–9, 94, 135–43, 143–6, 195, 219, 224–5, 227–8, 244, 256, 272, 274.
environmental 6, 10, 65, 79, 84, 99, 155, 225, 231, 239, 264–8, 269, 271, 273–5, 277.
erotic 109.
eternal 68, 237, 267.
ethnic 2, 5, 9, 81, 84, 86, 98, 224–5, 271.
evil 34, 57, 113–4, 135, 137, 138–40, 143–5, 150, 155, 159, 164, 272–3, 275.

exotic 86, 94, 258–9.
exploitation 56, 106, 108, 126, 141, 157.
fae 1, 4–11, 28, 34, 79–88, 104–6, 108, 136, 154–60, 195, 198, 216–8, 223, 225–8, 262, 266–7.
faerie 52, 63–5, 69, 79, 84, 85, 104, 105, 144, 153, 156, 216–8, 227–8, 231, 243–9, 257–63.
"Faerie Queene" (1590) 262.
faith 4, 47, 49–51, 223.
fairytale 3, 6, 27, 34, 39, 42, 45–6, 49–51, 73, 99, 103, 121, 123, 126–9, 153, 178, 216, 228, 235–6, 238, 256, 267.
fairy
— culture 10.
— dust 212.
— fiction 37–8, 40, 42.
— folk 3, 17–8, 171, 173.
— fort 74, 76, 170.
— Godmother 5, 8, 47–8, 51, 121–2, 124–9.
— land 5, 10, 77, 136, 131, 139, 154–8, 160, 217, 233–4, 343.
— like 3, 8, 9, 182, 243.
— lore 2, 4, 27, 28, 33, 42, 57, 59–61, 77, 168, 267.
— made 194.
— magic 57, 193, 237.
— people 4, 38.
— queen 59–60, 137, 267.
— realm 6, 27, 63, 71, 105, 244, 247, 274.
— struck 112, 173.
— trees 170–2, 252.
— tropes 104.
— world 2, 3, 5, 78, 223–8, 230, 277.
fake 21–2, 25.
fantasy 4–5, 6, 9, 38, 39, 43, 47, 51, 64, 73, 79, 135, 138–9, 140, 192, 207–9, 219, 223, 228, 230.
— science 42.
fantasyland 96–7.

Index

father 18, 55, 69, 84, 122–3, 126, 155.
— Hog 136.
— hood 8, 73–8.
feminine 7–8, 49, 75, 124, 127, 200.
femininity 48, 94, 122, 127, 148.
feminism 121–2, 124–7, 144, 146, 149–51, 179.
fertility 67, 70, 267.
fetish 66, 86.
flowers 7, 10, 13, 37, 39, 40, 43, 101, 135, 149, 163–4, 187, 194, 198, 219, 224, 226, 228, 234–5, 238, 262, 264–6.
folklore 1–4, 10, 28, 58, 61, 66–8, 74, 111–3, 118, 135–6, 143–5, 148, 150, 153, 155, 158, 160, 167–8, 175, 177, 181, 183, 216, 219, 228, 234, 243, 252–60, 262–3, 269, 271.
folkloresque 37–44, 251–3, 255, 257–60.

game 6, 10, 96, 177, 215–21, 255, 261–4, 268.
gender 2, 4–6, 8, 79, 86, 88, 96, 104, 107, 118, 121–3, 126–7, 129, 137–8, 144, 146–7, 154, 176, 178–80, 186, 190, 221, 228.
ghost 49, 68, 73, 154, 163, 181, 188, 115, 218, 243.
giant 10, 67, 219, 224, 244, 254, 263.
goblins 25, 42, 136, 143–6, 159, 223, 250.
God 129, 168, 244, 276.
goddess 47, 105, 164.
gods 59, 71, 86, 136, 168, 225, 231.
golem 139, 215, 219.
Gothic 64, 67, 69, 71, 78–9, 147, 215, 223.
green 86, 97, 194–5, 238–40, 244–5, 263, 266.
— ambassador of, 99.
— children 243–5, 250.
— eyes 164, 256, 264.
— man 42, 70, 71, 219.

halfling 159, 219.
half
— blood 84–5.
— human 74, 82.

hoax 22.
hobgoblin 42, 144.
homosexuality 220–1.
hybrid 103–4, 106, 108–9, 153, 160, 162.

identity 6, 10, 80, 84, 103–7, 117–8, 121, 124, 146, 156, 158, 160, 175, 179, 181, 186–7, 189, 218–21.
— Black 85.
— national 7, 63–4, 71.
— self 9, 109.
illness 28, 112, 173, 252.
India 9, 175, 188.
Indigenous 5, 83–4, 269, 271–4, 277.
infect 270.
invisible 10, 68, 170, 172, 186, 188, 233–6, 238, 240, 265, 272.
Ireland 4, 8, 30, 75, 111, 113–4, 118, 143, 155, 167–9, 173, 252, 254.
Irish 3, 9–10, 28, 30–2, 68, 73–7, 78, 82, 111–3, 117, 136, 167–73, 251–60, 261–3.

Japan 3, 9, 181–90, 195, 227.
jealousy 99, 101, 177.
jest 117, 210.
jester 139.
joke 22, 38, 117, 155.

King 9, 38, 56, 60, 63, 70–1, 137, 139, 140, 177–8, 247–9, 256.
— Arthur 66, 70.
kingdom 39, 54, 60, 70, 82, 83, 86, 87, 96, 100, 102, 247–8, 263, 272.

landscape 6, 9, 24, 47, 55, 63–71, 85, 116–7, 148, 156, 170, 172, 200, 252, 266, 269–77.
language 10, 24, 32, 111, 114, 136, 157, 169, 190, 195, 243, 258–9, 272, 274.
legend 2–4, 27–8, 32–3, 35, 56–7, 59–60, 70, 77–8, 140, 167, 181, 244, 246, 250.

liminal 50, 67–8, 71, 82, 104, 107, 146, 153–4, 158, 160, 201, 216, 247, 250.
London 9, 22, 75, 81, 113, 153–4, 157–60.

madness 34, 63, 67, 217, 220, 252.
mathematics 10, 207–14, 239.
magic 1, 3, 5–6, 11, 21, 34, 41, 50, 57, 63–5, 67, 68–71, 94–7, 99, 102–4, 109, 129, 135–6, 138–9, 140, 153–4, 156–60, 165, 168–9, 172, 182–4, 188, 192–3, 198, 201, 207–14, 215, 218–20, 231, 233–40, 251, 254–5, 257–8, 261, 263–4.
Magical Kingdom 96, 100.
male-dominated 75, 139.
masculine 271.
materialism 24–5, 268.
material world 23, 25, 65, 67.
maternal 112, 115–8, 121, 126–7.
matriarch 261–3, 265, 267.
Medieval 3, 10, 55–7, 65, 170, 224, 229–30, 240, 243–4, 246–8.
Melusine 56, 60.
memory 2, 42, 51, 65, 67–8, 83–4, 138, 158–9, 261, 264–6.
A Midsummer Night's Dream (1596) 3, 5.
monster 143–4.
monstrous 103, 139, 158.
moon 24, 83, 86, 87, 165, 238, 246.
— light 9, 48, 163–5.
mother 18, 28, 34, 48–9, 51, 112, 116–9, 126–7, 146, 155, 164, 178–9, 212.
— adoptive 115.
— fairy 55, 84.
— foster 77.
— grand 107.
— hood 8, 111–4, 116–9.
— nature 231.
— step 122, 126.
muse 7, 8, 91, 219, 223, 228.
music 10–1, 73, 124, 128, 163, 168, 171, 185–6, 188–9, 223–5, 227–8, 230–1, 246, 270.

mutation 2, 6, 200.
mystical 1, 42, 66, 220.
myth 7, 42, 45, 60, 66, 69–71, 78, 99, 112, 118, 163, 273.
mythical 215, 220.
mythological 4, 43, 55–8, 79, 82, 86, 136, 145, 223, 228, 244, 262–3, 267.

Native American 219
Native Australian 274. 276.
nature 9–10, 66–7, 83–4, 143–4, 181, 226, 228, 230–1, 233–6, 238, 240, 243, 247, 263, 266–7, 270, 273–7.
nightmare 43, 74, 137, 143, 164, 217, 247.
Northern Ireland 73.
nostalgia 66, 156, 225, 227, 242.

Otherness 7–8, 76, 79, 86, 88, 94, 115, 228.
otherworld 1, 9, 167–9, 173, 248, 250.
otherworldly 155, 157, 228, 244–6.
outsider 7–8, 114–5, 117, 159, 211, 259,

pagan 55, 60–1, 164, 223, 230, 244, 246, 250.
paradise 96, 246–48, 250, 263–4, 266–7.
pastoral 60, 266.
performative 57–8, 60, 178, 231.
Peter Pan 4, 8, 37, 40, 93–7, 99, 101, 194, 210.
photograph 17–8, 20–3, 26, 95, 98, 100, 225.
Pinocchio 45–9.
pixies 1, 4–5, 94, 144, 195–6, 276.
Pixieville 237.
politics 2, 6, 45, 56–8, 60–1, 79–88, 94, 137, 144, 146, 181, 208, 219, 265.
pookas 216.
possession 5, 77.
pregnancy 112, 114–6, 118.
princess 21–2, 93, 98–9, 122, 124, 207, 209.
Queen 56, 61, 107–8, 136, 139–40, 157, 177–9, 193, 236, 244, 246, 263, 265–6.

Index

— Elizabeth I 7.
— Mab 108, 262, 263.
— Oona 10, 261–2, 267.
Queene of Phyries 7, 55, 59.
queer 8, 126–7, 220–1.

race 2, 23, 67, 76, 79–80, 82, 84–5, 87–8, 112, 122, 129, 137, 153, 217–9, 252, 263–6, 271.
racism 81, 157.
religion 1–2, 33, 42, 45, 49–50, 58, 65, 111, 177, 183, 208, 225, 243–4, 246, 250.
resurrection 64, 67, 69, 248.
ritual 4, 13, 33, 176, 183, 248.
robot 45–6, 48–51, 239.
Romanticism 64, 67, 71, 143, 208, 224, 228, 230, 268.
rural 64, 75, 111, 113–4, 117–8, 155.

Samhain 159, 163.
Scandinavia 3, 219.
science 7, 10, 23–5, 65, 189, 233–40, 260, 266.
science fiction 43, 45, 51, 257.
Scotland 136, 143, 145, 155, 249, 252, 256.
seelie 256.
selkie 144, 216.
sexualised 5, 8, 48, 99, 191.
— hyper 86.
sexual 86, 94, 107, 154, 244.
sexuality 146,
sídhe 10, 259.
slavery 63, 70, 137, 154, 156–8, 160, 218.
spirit 18, 23, 25, 59, 143–51, 153, 159, 182–3, 185–6, 190, 209, 215, 219, 225, 230, 234, 243, 270–6.
spirituality 6, 81, 181, 183, 225, 274, 277.
— eco 269.
spiritualism 17, 22–3.
supernatural 1, 3, 7, 10, 25, 41–2, 45–6, 49–50, 55–7, 60, 65, 67, 73, 76, 78, 82, 104, 106, 108, 111–2, 116, 143–5, 148, 150, 153–4, 159, 169, 171–2, 188, 209, 215–6, 233, 243–4, 246–7, 252, 262.
superstition 33, 78, 111–5, 118, 145.
symbolic 10, 46, 48–9, 66, 69, 74, 76, 127, 244, 250.
symbolism 47–8, 61, 64, 70–1, 94, 104, 115, 123, 226, 239, 247.

technology 7, 24, 45–6, 97, 101–2, 176, 191–4, 200–1, 236, 238–42.
Tinker Bell 4–5, 8, 37, 40–1, 93–102, 168, 173, 193–4, 209–11.
Tolkien, J. R. R. 43, 137, 219, 227–8, 263.
tooth fairy 135–6.
tradition 3–6, 27–8, 37, 41, 43, 59, 67, 127, 140, 145, 150, 153–6, 167, 171, 173, 223, 225, 228, 230–1, 236, 249, 253, 274.
traditional 7, 9, 33, 43, 46, 95, 98, 103, 111, 114, 116–7, 123–4, 127, 141–4, 147, 168, 181, 193, 219, 227, 229, 234, 237, 254, 267.
transformation 3, 9, 125, 193, 195, 198–200, 209, 243, 265, 276.
trauma 76, 112–4, 116, 118, 126, 157–9, 218, 270.
treefolk 265–6.
trickster 1, 148, 265.
troll 135–6, 139, 144, 219.

Underworld 66, 246–8.
urban 9–10, 79, 81–2, 117, 153–4, 156–60, 184, 230, 251, 255, 257–8.
utopia 263–4, 266–7.

vagabond 160.
vampire 6, 85, 103–7, 109, 154, 192, 215.
Victorian 3, 7, 10, 23–4, 27–8, 41, 79, 81, 168, 195, 198, 207–8, 230, 233–4, 236, 243, 251, 262, 268.
violence 30–1, 33, 37, 40, 43, 61, 81, 112, 148, 177–8.

werewolves 135, 215.
whiteness 84–6, 88, 127–8, 136, 147, 191, 273.
whitewashing 273.
wings 1, 40–1, 74–5, 79, 84–7, 95, 97, 208, 213, 257.

Winx 5, 191–5, 197–9, 238–40.
witch 78, 136, 138, 140, 171.
woodland 66, 163, 234.

Genre Fiction and Film Companions

Series Editor: Simon Bacon

The *Genre Fiction and Film Companions* provide accessible introductions to key texts within the most popular genres of our time. Written by leading scholars in the field, brief essays on individual texts offer innovative ways of understanding, interpreting and reading the topics in question. Invaluable for students, teachers and fans alike, these surveys offer new insights into the most important literary works, films, music, events and more within genre fiction and film.

We welcome proposals for edited collections on new genres and topics. Please contact baconetti@googlemail.com or oxford@peterlang.com.

Published Volumes

The Gothic
Edited by Simon Bacon

Cli-Fi
Edited by Axel Goodbody and Adeline Johns-Putra

Horror
Edited by Simon Bacon

Sci-Fi
Edited by Jack Fennell

Monsters
Edited by Simon Bacon

Transmedia Cultures
Edited by Simon Bacon

Shirley Jackson
Edited by Kristopher Woofter

Toxic Cultures
Edited by Simon Bacon

Magic
Edited by Katharina Rein

The Undead in the 21st Century
Edited by Simon Bacon

The Deep
Edited by Marko Teodorski and Simon Bacon

Death in the 21st Century
Edited by Katarzyna Bronk-Bacon and Simon Bacon

Alice Through the Looking-Glass
Edited by Franziska E. Kohlt and Justine Houyaux

The Weird
Edited by Carl H. Sederholm and Kristopher Woofter

Aliens
Edited by Elana Gomel and Simon Bacon

IndigePop
Edited by Svetlana Seibel and Kati Dlaske

Fairies
Edited by Lorna Piatti-Farnell and Simon Bacon

www.ingramcontent.com/pod-product-compliance
Ingram Content Group UK Ltd.
Pitfield, Milton Keynes, MK11 3LW, UK
UKHW021254180426
11947UKWH00010B/766